AF839701

Development of Molecular Methods to Detect Infectious Viruses in Water

The mission of the Awwa Research Foundation (AwwaRF) is to advance the science of water to improve the quality of life. Funded primarily through annual subscription payments from over 1,000 utilities, consulting firms, and manufacturers in North America and abroad, AwwaRF sponsors research on all aspects of drinking water, including supply and resources, treatment, monitoring and analysis, distribution, management, and health effects.

From its headquarters in Denver, Colorado, the AwwaRF staff directs and supports the efforts of over 700 volunteers, who are the heart of the research program. These volunteers, serving on various boards and committees, use their expertise to select and monitor research studies to benefit the entire drinking water community.

Research findings are disseminated through a number of technology transfer activities, including research reports, conferences, videotape summaries, and periodicals.

Development of Molecular Methods to Detect Infectious Viruses in Water

Prepared by:
Theresa L. Cromeans and **Jothikumar Narayanan**
Department of Environmental Sciences and Engineering, School of Public Health, University of North Carolina, Chapel Hill, NC
and Division of Viral Hepatitis, NCID, CDC, Atlanta, GA
and
KwanHye Jung, **Gwangpyo Ko**, **Doug Wait**, and **Mark D. Sobsey**
Department of Environmental Sciences and Engineering, School of Public Health, University of North Carolina, Chapel Hill, NC

Sponsored by:
Awwa Research Foundation
6666 West Quincy Avenue, Denver, CO 80235-3098

Published by:
Awwa Research Foundation
Advancing the Science of Water®
IWA Publishing

DISCLAIMER

Published by IWA Publishing, Alliance House, 12 Caxton Street, London SW1H 0QS, UK

Telephone: +44 (0) 20 7654 5500; Fax: +44 (0) 20 7654 5555; Email: publications@iwap.co.uk;
Web: **www.iwapublishing.com**

AwwaRF report number 90995F. Originally published by AwwaRF for its subscribers in 2004.
IWA Publishing version published 2005.

ISBN 1 84339 891 5

CONTENTS

TABLES

FIGURES

FOREWORD

The Awwa Research Foundation is a nonprofit corporation that is dedicated to the implementation of a research effort to help utilities respond to regulatory requirements and traditional high-priority concerns of the industry. The research agenda is developed through a process of consultation with subscribers and drinking water working professionals. Under the umbrella of a Strategic Research Plan, the Research Advisory Council prioritizes the suggested projects based upon current and future needs, applicability, and past work; the recommendations are forwarded to the Board of Trustees for final selection. The foundation also sponsors research projects through the unsolicited proposal process; the Collaborative Research, Research Applications, and Tailored Collaboration programs; and various joint research efforts with organizations such as the U.S. Environmental Protection Agency, the U.S. Bureau of Reclamation, and the Association of California Water Agencies.

This publication is a result of one of these sponsored studies, and it is hoped that its findings will be applied in communities throughout the world. The following report serves not only as a means of communicating the results of the water industry's centralized research program, but also as a tool to enlist the further support of the nonmember utilities and individuals.

Projects are managed closely from their inception to the final report by the foundation's staff and large cadre of volunteers who willingly contribute their time and expertise. The foundation serves a planning and management function and awards contracts to other institutions such as water utilities, universities, and engineering firms. The funding for this research effort comes primarily from the Subscription Program, through which water utilities subscribe to the research program and make an annual payment proportionate to the volume of water they deliver and consultants and manufacturers subscribe based on their annual billings. The program offers a cost-effective and fair method for funding research in the public interest.

A broad spectrum of water supply issues is addressed by the foundation's research agenda: resources, treatment and operations, distribution and storage, water quality and analysis, toxicology, economics, and management. The ultimate purpose of the coordinated effort is to assist water suppliers to provide the highest possible quality of water economically and reliably. The true benefits are realized when the results are implemented at the utility level. The foundation's trustees are pleased to offer this publication as a contribution toward that end.

Edmund G. Archuleta. P.E.
Chair, Board of Trustees
Awwa Research Foundation

James F. Manwaring, P.E.
Executive Director
Awwa Research Foundation

ACKNOWLEDGMENTS

We thank Douglas A. Wait and Dorothy Thompson for all processing of water samples and other technical assistance. We thank Angela S. Foust for excellent technical work and organization. We thank Donna Simmons for contributions to document preparation.

We acknowledge the collaboration of Dr. Ralph Baric, Department of Epidemiology, School of Public Health, UNC. We acknowledge the collaboration of Dr. Dean Erdman, Dr. Steve Monroe and Dr. Steve Oberste, DVRD, NCID, CDC. We acknowledge the collaboration of Dr. Betty Robertson, DVH, NCID, CDC and the excellent facilities at CDC-Atlanta that facilitated a portion of this work.

We thank the utilities for their contributions and participation:

- Clayton County Water Authority, Morrow, GA
- Columbus Water Works, Columbus, GA
- City of LaGrange Public Utilities
- Cobb County Marietta Water Authority Wyckoff Treatment Division, Acworth, GA.

We thank our project officer Misha Hasan for patient guidance and communications. We thank our Project Advisory Committee:

- Dr. David Battigelli
- Dr. Steven Monroe
- Dr. Suresh Pillai
- Dr. Mitch Stoker

EXECUTIVE SUMMARY

There continues to be an urgent need to detect enteric viruses and other pathogens of public health concern in drinking water and other environmental samples by sensitive, specific and rapid methods. Many enteric viruses of concern in water are not culturable or are cultured only with difficulty and inefficiently in host cells. In the last decade molecular methods targeting nucleic acids have been developed and successfully applied for the detection of viruses in environmental samples. Although not routine tests, these molecular methods may become routine as technology improves and simple commercial kits become available. The most sensitive methods are those that specifically amplify nucleic acid sequences of the virus. For DNA viruses, the amplification method usually is polymerase chain reaction (PCR) and for RNA viruses, it is reverse transcription-(RT-) PCR. In theory, one nucleic acid target can be amplified by as much as a million-fold or more in this process, thereby providing the much-needed sensitivity to detect even the low levels of viruses sometimes found in drinking water and its sources. However, these methods detect only the viral nucleic acid, which is evidence of virus presence, not of virus infectivity.

In order for a virus particle to be infectious for the host, it must possess: (1) an intact outer coat (capsid) that is reactive with the appropriate cellular receptors and (2) a complete, functional viral genome capable of initiating infection (replication) within the host cell. Damage to either of these components will result in loss of infectivity (inactivation) of the virus particle. Viruses in natural waters may be either infectious or inactivated due to environmental exposure after release from the contamination source or due to inactivation by wastewater treatment processes prior to their release into natural water. While drinking water treatment of raw source water ideally results in removal or inactivation of infectious viruses, very low levels of viruses may survive.

In this study new molecular approaches for detection of infectious viruses were developed that are based on detection of both intact and functional virus capsids and intact and functional (replicating or transcribable) nucleic acids. These approaches are intended to prevent or minimize both false positive and false negative results associated with direct viral nucleic acid detection. In one form of the method, functional virus capsids react with host cellular receptors to mimic the first step of infection. Viruses are captured by their interaction with a soluble form of the specific cellular receptor of a group of viruses. Following the specific interaction with the receptor, nucleic acid of the captured virus can be released from within the virus particle and amplified by PCR or RT-PCR to provide the necessary detection sensitivity. Another approach developed in this study is the selection and capture of potentially infectious virus using specific antibody to the functional capsids of infectious viruses.

Detection and assay of infectious virus often is accomplished with conventional cell culture techniques. After inoculation of cell cultures with samples containing infectious viruses, cultures are observed for evidence of virus replication and infection as morphological changes called cytopathic effects (CPE) often characteristic of the virus type or group. However, some viruses infect cells and replicate without causing the CPE, making them difficult to detect unless special analytical methods are employed. One recently developed analytical method is to examine the inoculated cells for the presence of an increased quantity of viral genome nucleic acid by PCR or RT-PCR. However, the nucleic acids of damaged and non-infectious virus particles may be detected with this method if such material is present in the sample material inoculated into the cell cultures. Therefore, with conventional PCR or RT-PCR this necessitates cumbersome work to demonstrate evidence of increased viral nucleic acid amounts beyond the amount present in the inoculum. In this study methods were developed for detection of the production of new and unique viral nucleic acids within the infected cells. This provides more conclusive evidence of infectious virus in the samples inoculated.

The positive-stranded RNA viruses, which include many enteric viruses potentially found in drinking water supplies (enteroviruses, coxsackieviruses, Noroviruses [formerly Norwalk-like viruses], astroviruses and hepatitis A and E viruses), produce a negative, complimentary strand of RNA in the early phase of viral replication. This negative strand of RNA exists bound to newly formed positive strands of RNA to form a double-stranded replicative form (RF). Detection of the RF in cell cultures inoculated with samples containing viruses is an assay to conclusively demonstrate virus infectivity. In this study virus-specific RF detection was developed as an indicator of infectious model viruses and then applied to the detection of field isolated (wild-type) viruses in environmental water samples.

An important group of DNA viruses listed on the Candidate Contaminant List (CCL) of EPA is the adenovirus family (Ad). This family includes two types, 40 and 41, that cause gastroenteritis and other types that are also fecally shed (Ad 2, 5, and 31) and may be found in contaminated water. The adenoviruses are very resistant to wastewater treatment and to UV disinfection. These viruses form messenger RNA (mRNA) during replication in cells. The mRNA of adenoviruses is a nucleic acid target indicative of and specific for infectious virus that can be detected by RT-PCR. This detection method was successfully applied to the detection of infectious adenoviruses and it did not detect inactivated viruses.

Real-time quantitative PCR is a newer method that simultaneously amplifies, detects and quantifies nucleic acids in a single tube assay within a short time (within 90 minutes) and has been developed for clinical use for some viruses and many other pathogens. This technique eliminates the use of agarose gels and subsequent, separate nucleic acid hybridization, which are extra time-consuming steps for detection of PCR products. In real-time quantitative (RT-)PCR the specificity of hybridization is included in the amplification reaction as fluorescent-labeled dyes are used in the form of labeled primers or probes. In this study the usefulness of the method was demonstrated for detection of viruses in water sample concentrates in conjunction with the development of reliable nucleic acid extraction methods that provide RNA and DNA samples of suitable quality and quantity for the detection method.

RESEARCH OBJECTIVES

This research was funded by the American Water Works Association Research Foundation (AwwaRF) to develop and evaluate molecular methods for detection of potentially infectious viruses in water concentrates. This report presents the development and application of the following methods: (1) immuno- and host cell receptor-mediated capture systems for infectious viruses, (2) an integrated cell culture/RNA amplification (RT-PCR) method for either replicative form (RF) of positive-stranded RNA viruses or mRNA of DNA viruses, and (3) development of real-time nucleic acid amplification/detection methods with fluorogenic hybridization for rapid, quantitative and sensitive detection of hepatitis A virus (HAV), coxsackieviruses (using CVB3 as a model), Noroviruses, and Adenoviruses in water concentrates. The specific objectives of the study were the following:

1. Develop and evaluate sensitive, selective capture of infectious viruses in water concentrates.

a. Use a coxsackievirus-adenovirus cellular receptor protein (sCAR; reactive with both CVB3 and Ad41), produced in soluble recombinant form, to selectively bind only virus particles with intact binding epitopes for subsequent detection of viral nucleic acid by PCR or RT-PCR.

b. Use polyclonal (hyperimmune antisera to HAV) and monoclonal antibodies (murine antibodies to HAV and CVB3) for the recovery of intact, antigenically reactive virus particles followed by RT-PCR detection of viral nucleic acid.

2. Develop and evaluate cell culture combined with rapid, sensitive RT-PCR and PCR methods for detection of viable/ infectious waterborne viruses.

a. Use CVB3 and HAV as model viruses for the development of a sensitive and specific assay based on the molecular (RT-PCR) detection of negative viral RNA strands (existing as double stranded or replicative forms) in cell cultures

b. Develop and evaluate an integrated cell culture plus RT-PCR approach to detect infectious mRNA of adenoviruses (Ad) 2 and 41 as models in cell cultures, based on a conserved region of the viral genome expressed late in infection.

3. Develop methods for rapid sensitive and specific quantitation of detected viruses.

a. Develop one-tube, real-time RT-PCR methods for rapid detection of HAV, CVB3, and NV, by incorporating hybridization for specificity.

b. Develop one-tube, real-time PCR for rapid detection of adenoviruses types 2/5 and 40/41

4. Validate the performance of the methods for water sample concentrates that contain substances inhibitory to the molecular assays

5. Validate the ability of the developed methods to distinguish infectious from non-infectious/inactivated viruses by applying them to viruses inactivated by different disinfectants having different inactivation mechanisms, specifically free chlorine and low pressure UV radiation.

a. Inactivate viruses with typical doses of free chlorine in water of controlled quality (buffered, chlorine demand-free water) to determine the ability of the developed ligand capture methods to distinguish free chlorine-inactivated from infectious viruses

b. Inactivate viruses with typical doses of low pressure UV radiation in buffered, chlorine-demand-free water to determine the ability of the developed molecular detection methods to distinguish inactivated from infectious viruses

6. Collect, concentrate and archive raw and finished water samples (and corresponding water quality data) for subsequent virus analysis.

a. Collect, concentrate and archive 10L raw water samples for virus detection by the developed methods.

b. Collect, concentrate and archive 100L finished water samples for virus detection by the developed methods.

c. Analyze archived sample concentrates for enteroviruses, adenoviruses and Noroviruses using the newly developed methods

APPROACH

RT-PCR or PCR has the potential to yield ambiguous positive results for the presence of viruses in water by detecting non-infectious or inactivated viruses. Several investigators have demonstrated that the detection by PCR or RT-PCR does not always indicate the presence of infectious virus in environmental samples (Sobsey et al. 1998A; Ma et.al. 1994; Maier et al. 1995). To overcome this problem this study employed additional processing steps that depend on functional components of the virus needed for infection as an approach to detecting only infectious viruses. These methods were intended to preclude the detection of non-infectious viruses and increase the potential for detecting infectious viruses

Integrity and biochemical reactivity of the virus capsid is vital to virus infectivity because receptor-mediated attachment of capsid epitopes (binding sites) to specific cell receptors is the initial step of the process of infection. The specific binding of the virus to epitope-specific neutralizing antibody or to the cell receptor cannot occur if the necessary capsid components are not intact and reactive. The soluble host cell receptor and monoclonal antibodies employed to capture viruses in this study were shown to have neutralizing activity (the ability to prevent virus infection as evidence of their ability to react with the functional epitopes of intact, infectious viruses). In order to distinguish between infectious and non-infectious viruses, the functional activity of the viral capsid was evaluated by comparing PCR or RT-PCR titers of viruses captured with either antibody or the soluble cell receptor binding to viruses or virus nucleic acids assayed directly by PCR or RT-PCR. For viruses captured by antibody or cell receptor binding methods, the detection by RT-PCR or PCR indicates the presence of a nucleic acid within the functional capsid of a virus that is likely to be infectious. To further verify the ability of these capture methods to detect only infectious viruses, they were also applied to viruses inactivated by chemical (free chlorine disinfection) and physical (UV irradiation) methods.

In the process of virus infection through virus attachment via cellular receptors, the virus nucleic acid is released intracellularly and replication or amplification of the viral genome occurs. Detection of the unique intermediate forms of the viral nucleic acid present only during infection and not in virions confirms the process of replication. Virus infection of host cells may occur in the absence of macroscopically or microscopically visible evidence of infection such as CPE and it may occur at such low levels that infection can be confirmed only by detecting the replicative forms present in infected cells. The approaches to detect the replication of RNA and DNA viruses in this study were based on RT-PCR amplification of these unique intracellular RNA species, specifically the negative strand and positive strands in replicative forms of RNA viruses (coxsackieviruses and HAV) and mRNA of adenoviruses.

Different disinfectants inactivate viruses by specific mechanisms of action targeting different viral components. Either capsid or nucleic acid (or both) components of the virus may be altered during disinfection. We evaluated the ability of the newly developed methods to detect only infectious viruses by applying them to virus inactivated by either free chlorine or UV radiation. If the methods worked, they would not detect such inactivated viruses. Viruses inactivated by free chlorine were not detected and adenoviruses inactivated by high doses of UV irradiation also were not detected.

CONCLUSIONS

1. The method for detecting the replicative forms (RF) of infectious RNA viruses, specifically the enterovirus coxsackievirus B3 and HAV, in host cells was successful. Detection of RF of an enterovirus was possible for cell culture flasks infected with as few as 4 infectious units within 24 hours of inoculation. The method was specific for only infectious viruses. Viruses inactivated with free chlorine were not detected because they did not infect cells and produce the specific RF of RNA targeted by the detection method. The specific RF RNA

detected in the assay remains present and stable in the infected cells for up to one week in enterovirus infected cells. Although rapid (1-day) detection was possible with the wild-type viruses we evaluated, it is possible some viruses would not be detected as rapidly if replication were slower with a very low level inoculum. Because of the stability of the RF in infected cells, longer cultivation times in the absence of CPE are possible for virus detection by this method. Non-cytopathic HAV replication can also be detected with the method in a susceptible host cell line. HAV replication is very slow and usually takes weeks to months and repeated virus-cell passages to demonstrate replication, as shown previously by numerous investigators. The RF RT-PCR method demonstrated replication within two weeks of culture of one strain of wild-type virus with known infectivity for primates. An inoculum of between 10^2 and 10^5 primate infectious units was required to establish infectivity by the RF detection method at 2 weeks post infection. Considering the usual difficulty with wild type HAV cultivation, this is a promising finding. However, the extent to which this approach will be successful in detecting the low levels of viruses typically present in water samples was not investigated in this study and is a remaining research need.

2. In this study methods were developed to detect Adenovirus mRNA by RT-PCR in inoculated cells as an effective, sensitive and rapid method to detect low levels of infectious Ad in water sample concentrates. Because only infectious viruses can produce mRNA, this assay is specific for infectious viruses. The mRNA RT-PCR method clearly distinguished infectious adenoviruses from non-infectious adenoviruses that were inactivated by free chlorine and UV radiation disinfection. The assay sensitivity was as low as 5 IU of Adenovirus 41 in 3 days of culture. As few as 2 IU Ad2 and 10 IU Ad 41 in water sample concentrates were detectable in inoculated cell cultures by 7 days using this method. The developed mRNA RT-PCR method is particularly applicable to detection of infectious Adenoviruses 40 and 41, because these viruses do not produce clear CPE in cell cultures, even with prolonged incubation times of several weeks, but they do produce readily detectable mRNA in infected cells.

3. Coxsackie B viruses, laboratory and field-isolated, can be captured on magnetic beads coated with the soluble cellular receptor. Production of the soluble receptor (sCAR) was established with a yield adequate for hundreds of reactions per prepared batch. The method is specific for infectious viruses as demonstrated by the lack of detection of chlorine-disinfected viruses. As few as 1-10 infectious units of virus can be detected in the assay of a 100 μl sample.

4. Real-time quantitative RT-PCR or PCR can detect most of the human pathogenic viruses that may contaminate water including: enteroviruses, HAV, Noroviruses and Adenoviruses. More than one type of virus can be detected in a single assay. A method was developed for the simultaneous detection of HAV and enteroviruses with a sensitivity of 8 plaque forming units in 4.0 mL of source water sample concentrate. This method was used to validate virus presence in two water sample concentrates that were positive for enteroviruses with a standard RT-PCR assay. These samples were further validated for virus presence by nucleotide sequence analysis. The real-time assays for HAV and for CVB model enterovirus yielded a sensitivity (a detection limit) as low as 0.8 PFU. Primers and probe for real-time detection of Genogroup I Noroviruses were developed and demonstrated to detect all strains of the Genogroup in clinical specimens. Two different types of RNA extraction from stool samples were used, automatic and manual. Both provided quality samples for detection of viral RNA by real-time RT-PCR. An adenovirus real-time PCR was developed that can detect all Ad species that may contaminate water, including Ad 2, 5, 40 and 41, with a sensitivity of 0.1 infectious units of Ad2 and 0.01 IU of Ad41. The real-time PCR or RT-PCR assay can be obtained in 60-

90 minutes after nucleic acid extraction, thereby providing a rapid, sensitive and specific method for detection of multiple viruses in water.

5. An improved RNA extraction method was developed which permits testing of a larger volume of water sample concentrates for viruses. It is a simple column method that employs commercially available columns, does not require the use of organic solvents and can be done with only a low speed centrifugation. Ten-liter samples of source water were concentrated to 30 mL and RNA was extracted from four mL of this sample, in contrast to the usual 100 μl sample used for RNA extraction on commonly used mini columns. A buffer developed in this project was used in conjunction with a commercially available column to accomplish the increased recovery of RNA. Because the midi column commercial kit contains buffer designed to extract from only one mL of sample, the improved method applicable to 4-ml samples represents a four-fold increase in sample size for RNA recovery.

RECOMMENDATIONS AND FUTURE RESEARCH

The work reported in this study demonstrates that molecular methods can be developed to selectively detect only infectious viruses in water sample concentrates. Each method has been validated for the specific viruses studied. Further work is needed to expand these methods to other viruses or to combine the developed methods in ways permitting more rapid detection of only infectious viruses. Recommended areas of future development are:

1. Evaluation of different cell lines besides BGM for culture of enteric viruses in field samples. The RF detection method we describe for positive-strand RNA viruses is very specific and more rapid than appearance of viral CPE, however a limiting factor is the appropriate cell line(s) and appropriate virus adsorption conditions for successful infection. In this study it was demonstrated that one cell line was more susceptible than others for a single stool isolate of HAV. To further evaluate the developed methods a study is recommended for enteroviruses detection in sewage assumed to contain a wide variety of viruses, such as that from a large US city. Sample concentrates from sewage should be evaluated for virus infection in a variety of cell lines already considered to be good candidates for enteric virus cultivation.

2. The success of the sCAR receptor for capture of CVB in water sample concentrates creates the potential for developing capture systems for other viruses using their appropriate cellular receptor(s). Although the sCAR receptor method was evaluated for capture of infectious adenoviruses, the method developed for Coxsackie B viruses was not effective, apparently due to inadequate receptor concentration or unfavorable steric presence on the beads. Therefore, further research is needed to develop a successful sCAR system for adenovirus capture and PCR detection. The future development of an appropriate solid phase capture system coated with multiple cellular receptors for a variety of viruses will offer the ability to simultaneously capture a number of different enteric viruses in water sample concentrates.

3. While the real-time RT-PCR and PCR as developed does not distinguish infectious or non-infectious viruses, it can be applied following any of the methods that specifically capture infectious virus such as immuno- or receptor capture on magnetic beads. The real-time assay is a one-tube rapid assay with the specificity of hybridization included by use of a hybridization probe.

4. The real-time assay also can be used to detect virus replication in cell culture. The quantitative aspect of the assay provides an easy method to confirm an increase in viral genomes over that found in the water sample concentrate inoculum. Use of the real-time assay for

integrated cell-culture RT-PCR and PCR should be further explored and developed for rapid, quantitative detection of a variety of infectious enteric viruses in water.

5. Additional research is needed to evaluate the methods for recovery and detection of infectious viruses by applying them to viruses that have been inactivated by agents other than free chlorine or low pressure UV radiation. Viruses inactivated by medium pressure UV disinfection, chloramines, chlorine dioxide, ozone and the proteolytic enzymes associated with microbes in natural waters and biological treatment processes (both water and wastewater) also need to be evaluated for their detectability by the developed methods.

CHAPTER 1
INTRODUCTION

GENERAL APPROACH TO THE DETECTION OF INFECTIOUS VIRUSES

Surface and ground water sources and subsequently even drinking water supplies may be contaminated by a number of different viruses that are human pathogens. These pathogens may originate from a variety of sources, including treated wastewater effluents and non-point source discharges, incidental release of untreated wastewater, or even primary contact recreational use. Enteric viruses have been detected in environmental waters and potable water supplies and have caused waterborne outbreaks (Gantzer et al, 1998; Divizia et al., 1998; Green and Lewis, 1995). Water supply safety is evaluated with methods for the detection of bacteria that are not accurate surrogates for the presence of viruses. Viruses that may be transmitted in water are generally more stable in the environment and more resistant to water treatment processes than bacteria. Infections with most viral pathogens require lower infectious doses than are required for infection with most bacterial pathogens. Even low levels of viral pathogens in water pose a public health risk.

Current methods for the detection of viruses in water samples include standard techniques for in vitro cultivation in cell cultures and the newer molecular methods for the detection of viral nucleic acid. The newer molecular methods are potentially more sensitive than standard methods for culture of viruses and they can be completed rapidly. However, molecular methods that rely on detection of the viral nucleic acid alone do not yield any information about infectivity of the viral pathogens. In this study we have developed methods for the detection of potentially infectious viruses by targeting particular critical aspects of virus infectivity, and viral components produced only during virus infection of cells.

The first step in infection with viruses is binding to the plasma membrane of the target cell and this interaction occurs by the binding to a specific receptor as shown in Figure 1.1. A coreceptor such as a signaling receptor may be involved in the actual entry process but the specific binding is mediated through the attachment to a surface glycoprotein as shown. The non-enveloped (non-lipid containing) viruses, which include all viruses transmitted by water routes, have common or similar modes of cellular entry, although the specific chemical properties of the sites to which the virus binds for entry can differ, depending on the virus or virus group.

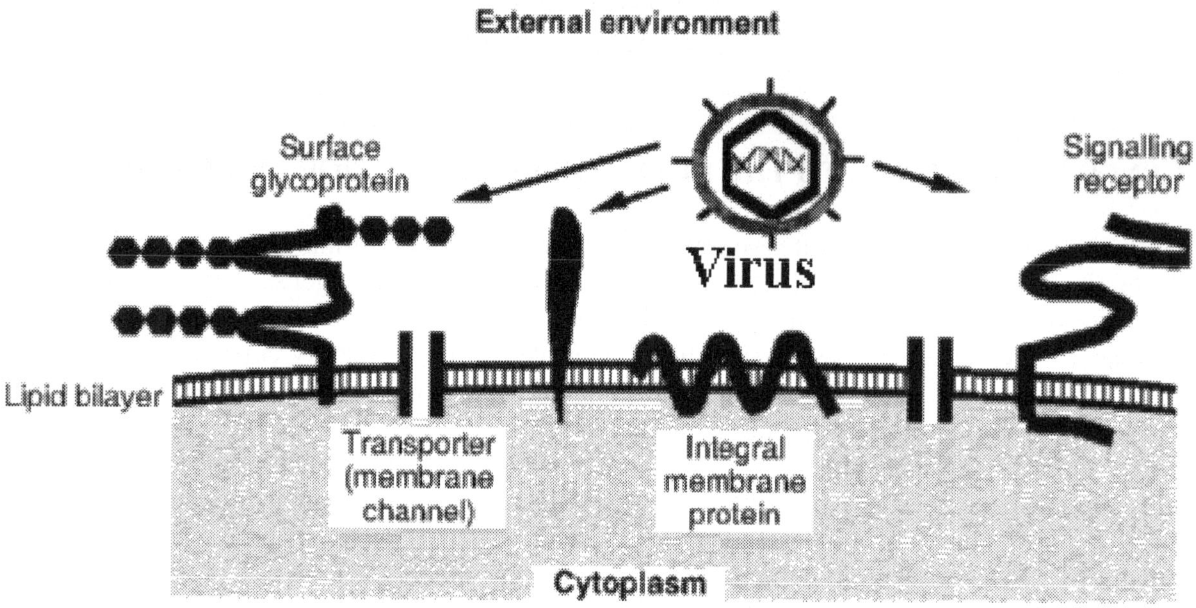

Figure 1.1. Diagram of cellular receptor (surface glycoprotein) for virus attachment to a cell. Adapted from All the Virology on the Web (2001) http://www.virology.net/

After or during entry into a cell the virus outer protein coat is removed and then the nucleic acids are released for viral replication to occur. Common features of the complete process from entry to production of new virus particles are depicted in Figure 1.2 for the positive sense RNA viruses. This class of viruses includes most viruses that may contaminate water, such as enteroviruses, caliciviruses, astroviruses and hepatitis A and E viruses. They are called positive sense since the viral genome serves as mRNA for production of the viral proteins. A negative strand complimentary to the positive strand is made. The negative strands only exist when replication occurs, but positive strands may be present from the original sample. Complimentary negative and positive strands exist in double-stranded forms, replicative forms (RF).

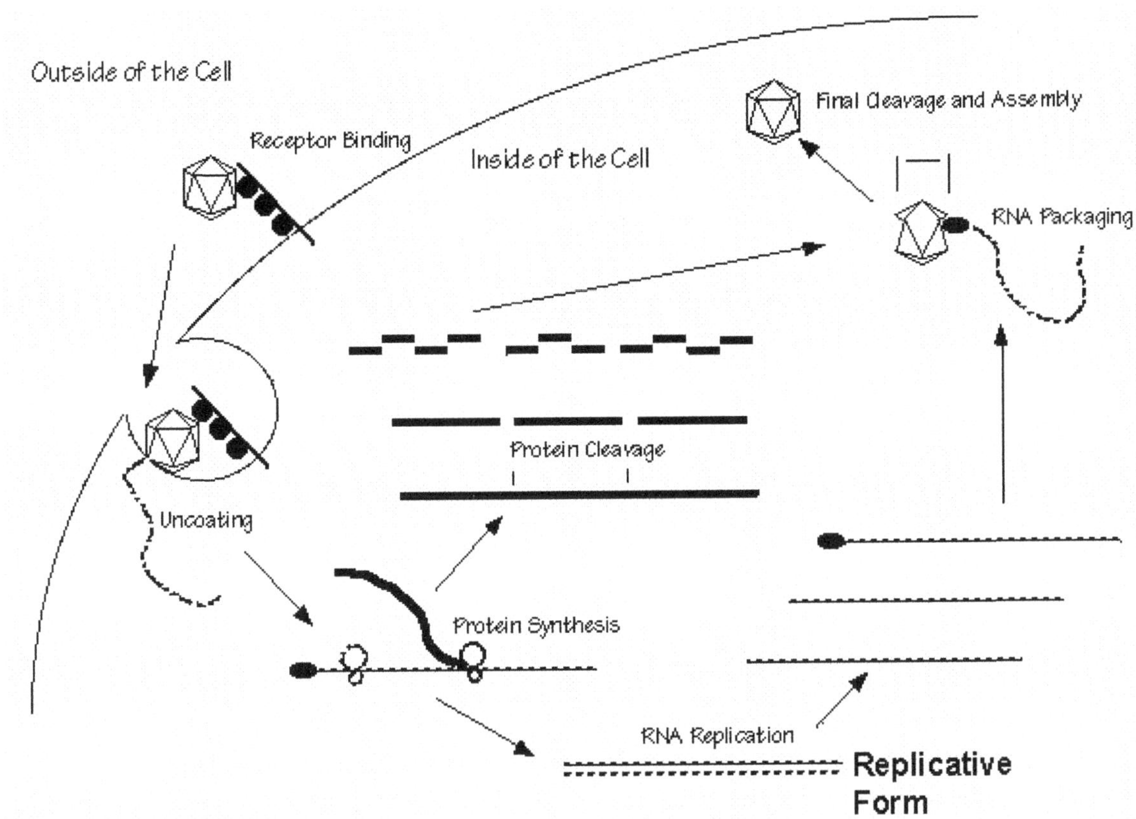

Figure 1.2. Diagram of the replication steps of positive-sense RNA viruses. Adapted from All the Virology on the Web (2003) http://www.virology.net/

Adenoviruses, an important family of viruses on the Candidate Contaminant List (CCL) identified by EPA cannot be detected by the RF method since they are double-stranded DNA containing viruses and have a different method of replication. The scheme of replication of the double-stranded DNA viruses is shown in simplified form in Figure 1.3. From the double stranded viral DNA, a series of messenger RNA (mRNA) molecules are made. The mRNA is functionally analogous to the RF of single-stranded RNA viruses and is only present when virus infection has occurred. The mRNA produces early proteins followed by late proteins during infection. Genetic regions for mRNA targets of both early and late proteins have been used in the evaluation of methods to detect infectious adenoviruses. A method was developed to detect only infectious adenoviruses with an RT-PCR specific for the newly made mRNA.

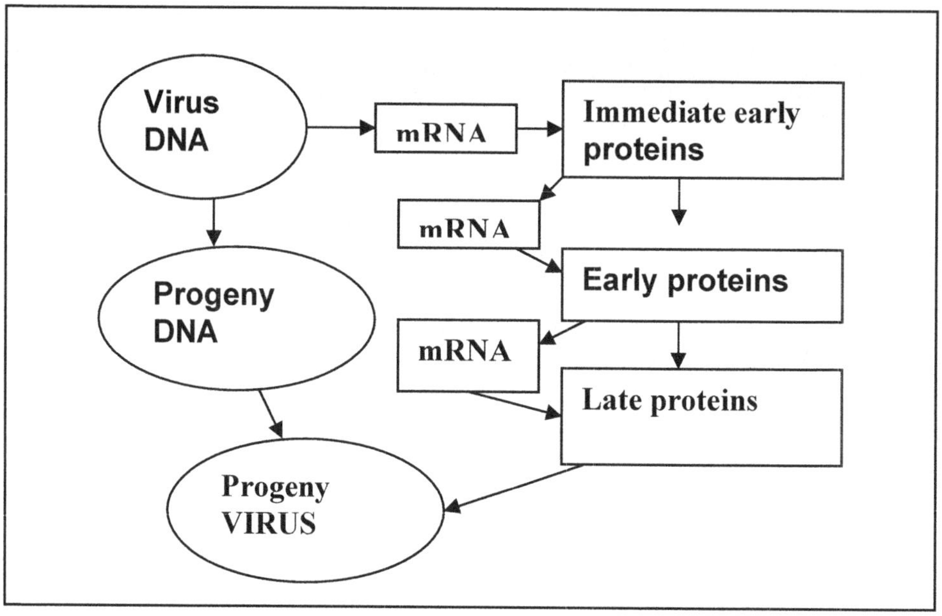

Figure 1.3. Schematic of replication of double-stranded DNA viruses including Adenoviruses. Adapted from All the Virology on the Web (2001) http://www.virology.net/

A method was developed for more rapid detection of the nucleic acid of viruses from water samples than the currently used standard PCR or RT-PCR methodologies. The standard PCR or RT-PCR methods require the additional and time-consuming steps of hybridization with molecular probes to confirm the presence of the correctly amplified nucleic acid. Real-time quantitative (RT-)PCR methods are those that confirm and identify the amplicons within the nucleic acid amplification assay, and can be completed quickly, perhaps in as few as 60 minutes. Although real-time RT-PCR does not directly detect infectious viruses, the technology can be applied to the detection of infectious viruses.

RECEPTOR-MEDIATED CAPTURE AND MOLECULAR DETECTION OF COXSACKIEVIRUSES AND ADENOVIRUSES

As previously described in Figure 1.1, the first step in virus infection of a host cell is attachment to a cellular receptor molecule. Attachment to the receptor then leads to the entry of the virus or its nucleic acid into the interior of the cell. These membrane-associated molecules are usually plasma membrane glycoproteins that usually serve some other critical cellular functions besides the initial site of virus infection (Rossman, He, and Kuhn, 2002). Binding of the virus usually destabilizes its structure and initiates the uncoating process by which the virus nucleic acid is released into the interior of the cell (Racaniello, 1996). The susceptibility of a cell to virus attachment and infection is mediated by the specific chemical composition of the cell's receptors. In principle, it should be possible to use the specific cell receptor of a virus or virus group as a virus binding or capture agent to recover intact and functional viruses from a sample. In this study this approach has been applied to the detection of coxsackieviruses in water.

The cellular surface protein that functions as a receptor for the six types of coxsackievirus B viruses as well as adenoviruses A, C, D, E, and F (40/41), called coxsackievirus-adenovirus receptor or CAR, was used in this study to capture infectious viruses (Bergelson et al. 1997). The receptor was produced in soluble recombinant form (sCAR) (Roelvink et al, 1998). The purified sCAR protein was shown to bind directly to viral capsids, and preincubation of virus with sCAR prevented binding to the cellular receptors on cultured cells, thereby preventing virus infection. sCAR is a 45-kDa-membrane glycoprotein with two Ig-like extracellular domains (D1 and D2), a transmembrane domain and a cytoplasmic domain. The role of sCAR binding in CVB and adenovirus infection is different. For CVBs, sCAR can function for both attachment and infection (entry of the viral nucleic acid into the host cell). For adenoviruses, the major function of sCAR is mediation of initial attachment, but subsequent entry of virus into the host cell is mediated by another molecule, integrin protein. The two different virus families use different binding sites on the D1 extracellular domain of sCAR.

The sCAR protein was produced in sufficient quantity by a baculovirus vector system in which it was expressed as an extracellular sCAR domain of 36kDa. A special protein tag included in the expressed sCAR protein, called FLAG, provided a mechanism for sCAR purification using an immunoaffinity gel chromatography column. After production of the purified sCAR protein, it was chemically bound to paramagnetic beads by a specific coating mechanism to assure proper orientation of the protein for virus binding and capture. The general process from production of sCAR protein to detection by RT-PCR of CVB is shown in Figure 1.4.

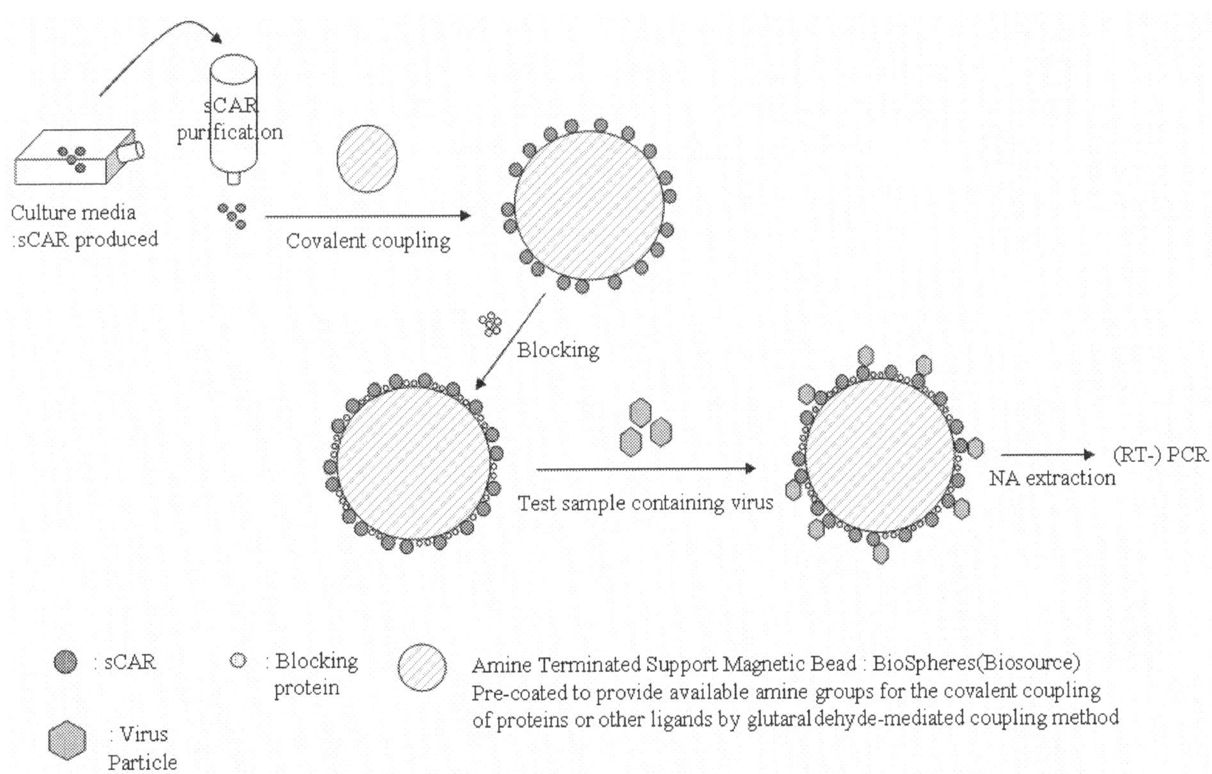

Figure 1.4. Schematic of the production of sCAR, binding to magnetic beads, capture of CVB, and detection by RT-PCR

The soluble cell receptor is a powerful tool for the detection of only those intact and functional viruses that are capable of binding and attaching to it. Viruses having capsids damaged by processes such as water treatment or by environmental stressors should not bind to the receptors. Following the ligand-specific capture of the virus particle by binding to sCAR, the nucleic acid amplification steps of the recovery and detection procedure provide the required sensitivity for detection of low levels of viruses in water samples.

DETECTION OF REPLICATIVE FORMS OF POSITIVE-SENSE RNA VIRUSES IN INFECTED CELLS

The standard method of assay and detection of infectious viral pathogens in environmental samples involves virus propagation in mammalian cell culture. The process of virus infection of host cells such as cell cultures involves a series of events that sometimes can be detected by the degeneration and death of the infected cells as a result of virus infection and replication. Many enteric viruses that pose a human health risk through waterborne exposure contain single-stranded positive-sense RNA genomes, such as enteroviruses, noroviruses, astroviruses and hepatitis A and E viruses. Replication of the viruses at the cellular level has some common features. The uncoated positive sense virus genomes encode for viral proteins, and complementary negative strands of RNA are formed from the positive sense genome. From the negative strands, positive strands are generated, and these are subsequently packaged in the new viral capsids to form viral progeny or used as messenger RNA to produce viral proteins. As replication proceeds, the positive strands are in excess. The negative strands exist in infected cells almost exclusively as a partially single stranded, partially double stranded replicative intermediate RNA during production of positive strands (Figure 1.5). After treatment with RNase A/T1, only the double stranded structures of positive and negative strand in equal quantity remain. Separate reactions with a dilution series of a sample will yield a semi-quantitative analysis of the negative strand concentration and the positive strand concentration in the preparation. If only positive viral strands are present in the cell, they will be digested and not detected.

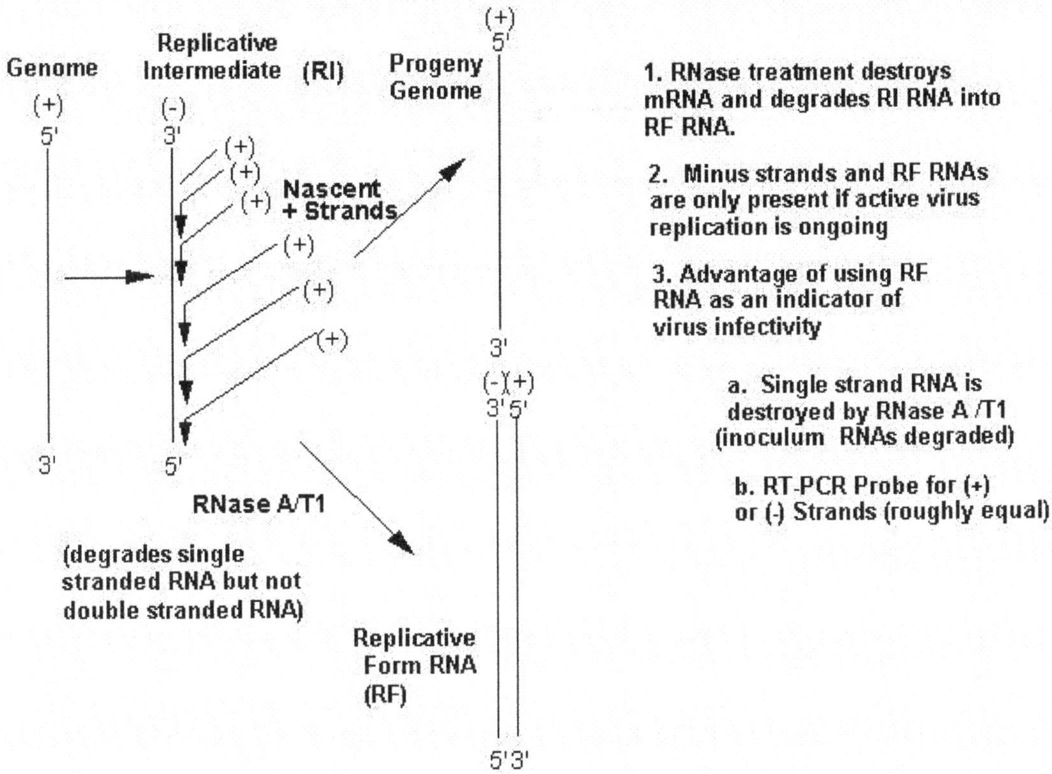

Figure 1.5. Replication of positive strand viral genomes and formation of replicative forms

Detection of the negative strand only could also confirm replication, however, there is a possibility of false positive results when relying only on the detection of negative strand. Despite this false-positive risk, negative strand detection for clinical assessment of viral replication in human tissues of hepatitis C and G viruses, coxsackieviruses, and polioviruses has been reported by several investigators (Lanford et al. 1995; Laskus et al. 1998; Andreoletti et al. 1997; Krawczynski et al. 1999; Reetoo et al. 2000; Cunningham et al. 1990; Novak and Kirkegaard, 1991; Chaves et al. 1994). Non-specific viral RNA detection was an issue in several of these studies and only recent studies have described multiple approaches to assure target specificity for single negative strand detection (Craggs et al. 2001).

Knowledge of the RF formation kinetics in enterovirus-infected cells is limited primarily to knowledge of the process based on experimental infection at a very high multiplicity of infection (MOI) with poliovirus in HeLa cells and other host cells (Johnson and Sarnow 1995; Barton and Flanegan 1997). Little is known about the kinetics of RF production during infection of cells with a very low quantity of virus, as typically found in most environmental samples. In this study a method was developed and evaluated for detection of RF as evidence of replication using low inocula levels of different enteroviruses.

The RF detection method also could be useful for viruses that are slow to replicate and do not cause CPE, such as hepatitis A virus (HAV). Because HAV is more stable in the environment and potentially more resistant to disinfection and environmental degradation than other enteric viruses (Cromeans et al, 2001), it is a good candidate for evaluation of RF production.

DETECTION OF INFECTIOUS DNA VIRUSES: ADENOVIRUSES

A similar approach for the detection of infectious DNA viruses as was taken for infectious RNA viruses in infected cells is examining the inoculated cells for presence of messenger RNA (mRNA). The role of mRNA in adenovirus (Ad) replication was described in Figure 1.3. This is an RNA species produced by the virus only during replication that is not found in the virus particle and hence not found in the water sample concentrates used as inocula.

Fifty-one different serotypes of Ads have been identified on the basis of neutralization with type-specific animal antisera, and these serotypes can be classified into six species (A-F) based on their ability to agglutinate red blood cells (DeJong et al. 1999). These classifications partially correlate with organ tropism of infection: respiratory diseases caused by subgroup B, C, E, epidemic keratoconjunctivitis by subgroup D, and gastroenteritis by subgroup F. Adenovirus types 40 (Ad40) and 41 (Ad41) in species F are potentially important waterborne viruses, demonstrate some resistance to wastewater treatment processes, and are known to be the second most commonly identified agent of gastroenteritis in children next to rotavirus (Brandt et al. 1985, Gerba et al., 2002).

Although Ad has been cultured in several cell lines, enteric Ad40 and Ad41 are difficult and slow to culture and do not produce clear cytopathogenic effects (CPE) in cell cultures (Tiemessen et al., 1996). The use of molecular methods targeting nucleic acids is increasingly useful for the detection of viruses and other waterborne microorganisms, including Ad. A number of studies have applied PCR for detecting Ad in various environmental settings such as swimming pools (Papetropoulou, and Vantarakis 1998), coastal waters (Jiang et al., 2001), river waters (Castignolles et al. 1998), sewage and shellfish (Pina et al. 1998).

However, direct detection of Ad DNA by PCR indicates only the presence of viral DNA and does not provide any information on infectivity, which is directly related to human health risk. Microorganisms can be inactivated by natural (sunlight, high temperature and other environmental stresses) or engineered (e.g., chemical disinfectants or UV radiation) processes, but their genomes can continue to persist over a relatively long period of time in the environment (Sobsey et al. 1998A). Furthermore, Ad has a high non-infectious particle to infectious unit ratio providing a significant population of non-infectious particles (Brown et al., 1992). The need to detect a few infectious viruses among many non-infectious viruses in environmental samples is important to predictions of public health risk in water and other environmental samples.

Integrated cell culture-PCR was developed to detect culturable viruses (Chapron et al. 2000; Chung et al., 1996; and Reynolds et al., 1996). This method combines cell culture and molecular detection of viral genomic nucleic acid. Cell culture prior to nucleic acid amplification increases the copy number of infectious viral nucleic acids, which leads to higher sensitivity and increased probability of detecting them, even if they do not produce CPE. However, as previously noted, this method has the potential to detect the nucleic acid of inactivated viruses that may have been in the sample that was inoculated into cell cultures. Such "carry-over" detection of nucleic acid of inactivated viruses could give a false positive result from samples containing no infectious viruses. However, mRNA can only be detected when infectious Ads enter cells and replicate. Therefore, detection of viral mRNA during cultivation is definitive indication of the presence of infectious Ads. We have developed a new molecular method to detect viable Ad based on RT-PCR of mRNA from cell culture.

The specificity of the molecular methods for infectious virus detection can be determined by evaluation of viruses treated with free chlorine or UV. These studies, which can be done in laboratory-scale, batch disinfection experiments using seeded, chlorine-demand-free buffered, reagent water and methods previously described (Battigelli et al. 1993; Shin and Sobsey 1998; Sobsey et al. 1991; 1998B), provide "benchmark" data on efficacy of these disinfection technologies in production of virologically "safe" drinking water.

Different disinfectants inactivate viruses by specific mechanisms of action targeting different viral components. Either (or both) capsid or nucleic acid components of the virus may be altered during disinfection. The mechanism of free chlorine disinfection for viruses has not been clearly established and varies according to virus type, suspending medium, pH, and dose (Thurman and Gerba, 1988). Standard water treatment doses of free chlorine probably affect both viral capsid and nucleic acid. Studies have reported damage to both capsid and nucleic acid depending upon the virus and conditions of the study (Alvarez and O'Brien 1982; Harakeh and Butler 1984). Inactivating doses of ultraviolet radiation primarily act upon the viral nucleic acids, by formation of pyrimidine dimers. The virus capsid is not appreciably altered unless UV doses are very high (higher than typically used in drinking water treatment; typically, 30-50 mJoules/cm^2). In this study the detection of mRNA of adenoviruses in cell cultures by RT-PCR was used to quantify infectious viruses. The cell culture-mRNA RT-PCR method was also evaluated for its ability to distinguish infectious viruses from viruses rendered non-infectious by disinfection with free chlorine or UV radiation.

REAL-TIME RT-PCR AND PCR FOR DETECTION OF VIRAL PATHOGENS IN WATER CONCENTRATES

There are some disadvantages of the current molecular techniques for direct detection of viruses in water via nucleic acid amplification. These methods are more sensitive and yield more specific information on the presence of viral pathogens than the traditional microbial indicator approach to assessment of water quality. However, standard RT-PCR and PCR require at least one day, usually more, for the nucleic acid amplification reactions and the subsequent confirmatory nucleic acid hybridization assays. Multiple sample tubes and repeated handling steps are required, providing opportunities for laboratory contamination and making the process labor intensive. Results are qualitative, and quantitative results can only be achieved with labor-intensive use of nucleic acid amplification of multiple sample dilutions. A rapid, quantitative nucleic acid amplification and identification method for specific detection of waterborne viruses is needed in order to provide greater opportunity for a quick response and implementation of control measures.

Recently, other investigators have reported the use of real-time RT-PCR for the detection and quantification of enteroviruses (Monpeoho et al. 2000; Monpoeho et al. 2001; Costa-Mattioli et al. 2002; Read et al. 2001; Verstrepen et al. 2001). Because multiple viruses may be found in a fecally contaminated water sample concentrate, detection of more than one virus in a sample would be most useful for water quality assessment. Multiplex RT-PCR, which allows for the simultaneous amplification of two or more different viruses in the same reaction tube, has been reported and successfully employed in the detection of viruses in environmental samples (Jothikumar, N. et al. 2000; Rosenfield and Jaykus 1999).

The real time RT-PCR assay is based on detection of an increase in fluorescence that is proportional to the amount of amplified product as it is generated during the PCR. There are as many as five main types of quantitative real-time (RT-)PCR chemistries currently described (Mackay et al. 2002). In this study, two general approaches were used. In one, a dye such as SYBR Green is bound to the newly formed double stranded DNA product and fluorescence increases as product quantity increases. With the use of the double-stranded DNA binding dye, the identity of the amplicons can be confirmed by the specific melting temperature profiles of the double-stranded DNA hybrids. The amplified DNA products also can be visualized with traditional gel electrophoresis in addition to the SYBR Green detection in real-time.

In the other real-time approach of this study, a labeled probe is used and real-time measurement of PCR product is the result of an increase in reporter fluorescence as the probe is displaced and cleaved by the 5'-3'-exonuclease activity of the Taq polymerase. The second step offers the specificity of the probe hybridization to specific sequences in the DNA synthesized by PCR amplification. The basic scheme is of this approach is shown in Figure 1.6.

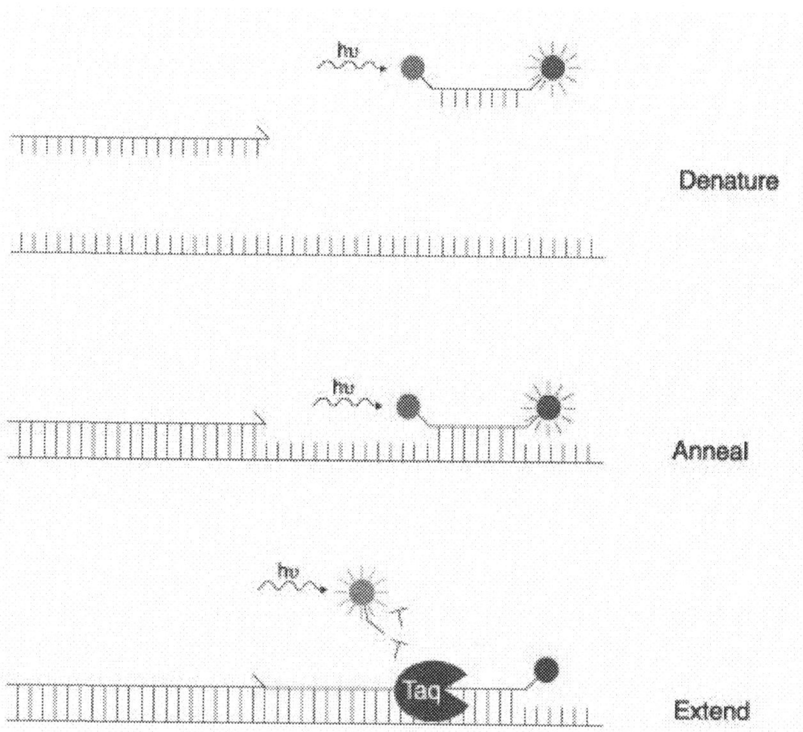

Figure 1.6. Schematic representation of real-time PCR with TaqMan labeled probe. Copied from Handbook of Fluorescent Probes and Research Products produced by Molecular Probes, copyright owned by Molecular Probes, Inc.

When the TaqMan probe is intact, energy is transferred via fluorescence resonance energy transfer (FRET) from the short-wavelength fluorophore on the left site (small circle on probe at top of diagram) to the long wavelength fluorophore on the other end (circle with radiant spikes), thereby quenching the short-wavelength fluorescence. When hybridized, the probe is susceptible to degradation by the endonuclease activity of a processing Taq polymerase. When degredation occurs, FRET is interrupted, increasing the fluorescence from the short-wavelength fluorophore and decreasing the fluorescence from the long-wavelength fluorophore. With repeated cycles of PCR, fluorescence increases and can be measured. The threshold or crossing point can be measured, which is the minimum number of cycles required to produce measured fluorescence in the exponential or log-linear region of the relationship between log concentration of target versus number of amplification cycles and which is considered the best estimator of target concentration.

Other than the reports of Monpoeho and coworkers (2000 and 2001) no detection of viruses in water concentrates by real-time RT-PCR or PCR has been published. In this study real-time, quantitative (RT-)PCR assays were developed and evaluated for enteroviruses, HAV, genogroup 1 NV, and Ad 2/5 and Ad 40/41. Primers and probes were specifically developed for real-time assays. A duplex real-time assay for enteroviruses and HAV also was developed.

No real-time RT-PCR assays for Noroviruses (NV) have been described in the published literature, primarily due to the limited information and perhaps lack of availability of known conserved regions of nucleotide sequences in the NV genome. The NV genus of the family *Caliciviridiae* has been divided into 4 genogroups based on the nucleotide sequences of a major genomic region encoding either the virus polymerase gene or the virus capsid. Genogroups I and II cause most identified human infections. The choice of primers and probe is limited by the genetic variation of the genomic regions for nucleic acid amplification and the requirement for detection of potentially low virus concentrations (low copy number of viral genomes) in water samples. Using available nucleotide sequence data of GI and GII Noroviruses available from clinical samples, a real-time, quantitative NV genogroup I detection system with specific primers and probes was developed in this study.

The developed duplex real-time RT-PCR system for enteroviruses and HAV was evaluated for recovery and detection in spiked surface water sample concentrates. This type of evaluation is important for validating the usefulness of the method for virus detection in field samples. One of the major problems associated with the detection of viruses in water sample concentrates by RT-PCR or PCR is the presence of amplification inhibitors in the samples, such as humic acids and polyphenols and heavy metals. These chemicals are likely to inhibit the *Taq* polymerase used to amplify nucleic acids by PCR (Ijzerman et al. 1997; Schwab et al. 1995; Staub et al. 1995). Although previously described procedures can remove inhibitors, they are relatively complex and can cause appreciable losses of target viruses or their nucleic acids. Many of the commercially available kits for nucleic acid purification before (RT-)PCR use phenol/chloroform extractions or column based purification after lysis with guanidine salts. The column-based methods are simple and use less hazardous chemicals, however, processing of large volumes of water sample concentrate is difficult and expensive, requiring the extraction of multiple small aliquots of the sample. The volume of the water sample concentrate is a limiting factor in detection sensitivity. Viral nucleic acid extraction and analysis of a larger volume of sample provides additional sensitivity (improved lower limit of detection). In the present study, a lysis buffer that can be used in conjunction with silica based spin columns was developed to extract RNA from up to 4 mL of a surface water sample concentrate with high sensitivity and speed. The method is compatible with real-time PCR and RT-PCR in addition to standard PCR and RT-PCR methods and provides effective removal of inhibitors in the process.

A TaqMan real-time PCR assay also was developed to detect and quantitate species C, D and F adenovirus DNA. The assay was evaluated and validated by correctly identifying the species of 25 recognized Ad prototype strains. For detection of species C, D and F, primers and probe were designed in the hexon gene, and primers and probe were designed from the fiber gene region for specific detection of Ad 40 and Ad 41. This assay was found to have a sensitivity (lower detection limit) of one infectious unit, and it detects all of the Ad species potentially found in contaminated water.

CHAPTER 2
METHODS AND MATERIALS

SOURCE AND FINISHED WATER COLLECTION AND CONCENTRATION

Finished water and source water samples were collected by four different water utilities and included a total of six different surface water sources and the finished water obtained from each source. Sampling was performed over a 22-month period on a monthly basis. The collection generated 104 samples of source water and 104 of finished water, which were collected, processed and stored for future evaluation.

Filter holders loaded with 142 mm Virosorb1 MDS filter disks and fitted with tubing connections and flow meters (filter set-ups) were shipped to each site monthly for collection of 100-liter sample volumes of finished water. Upon arrival at the UNC laboratory, these filters were sequentially eluted with 25 mL each of (1) 0.5 M lysine, pH 8.5, (2) 0.5 M arginine, pH 8.5, and (3) 3% beef extract (BE) + 0.05 M glycine pH 9.5. Lysine and arginine eluates were combined and concentrated to 1-2 mL volumes by polyethylene glycol (PEG) precipitation. After addition of 0.3 M NaCl with stirring to the eluates, 8% PEG (8000 MW, Sigma Co., Saint Louis, MO) was added and dissolved with magnetic stirring. Samples were incubated overnight at 4°C. Concentrates obtained by centrifugation at 7,000 $\times g$ for 30 min at 4°C were resuspended in one mL of Dulbecco's PBS (DPBS). The BE eluate was separately concentrated by PEG precipitation using similar methods. Concentrates were stored at -70^{0} C until thawed for testing.

A ten-liter volume from each source water was collected and shipped to the UNC laboratory with delivery in 24 hrs. Samples were processed upon arrival. Viruses were precipitated from the 10-liter samples of source water by PEG precipitation with 0.3 M NaCl and 8% (w/v) PEG (8000 MW) as described for drinking water samples. Each 10-liter sample was concentrated to about 30-40 mL in DPBS. Each sample was divided into 3 equal aliquots and stored at – 70 °C until one aliquot was thawed for testing.

Preparation of water sample concentrates for spiked sample model experiments was performed to replicate the volume and concentration conditions of the field sample collections at utilities. River water samples collected from the Chattahoochee River near the Recreation Park, Atlanta, GA, were concentrated with the same method used for the field samples. Samples from a small lake and pond were collected near Beaufort, NC, and samples collected from a larger lake near Chapel Hill, NC, were also processed by the same methods.

LABORATORY PROPAGATED CELLS AND VIRUSES

Hepatitis A virus and Coxsackievirus B3

CVB3 (Nancy strain, ATCC VR30) was passaged two times in BGM cells. The virus was harvested by freeze-thaw of cells three times followed by chloroform extraction. The aqueous supernatant after chloroform extraction was the stock virus, and it had a titer of 2 X 10^7 PFU/mL in an assay of serial ten-fold dilutions on BGM cells. Experimental work with CVB3 was performed with BGM cells (CDC stock originally from Biowhittaker).

HAV, HM-175 clone 24A, was propagated in FRhK-4 cells as previously described (Cromeans et al. 1989). A large stock of passage level 4 of this cloned stock was prepared. Cells were harvested at one-week post infection with 3 freeze-thaw cycles of cells and medium. This material was sonicated for 2 min. in a water bath sonicator (Sonic dismembranator 60, Fisher Scientific) after which it was extracted with trichlorotrifluoroethane. The upper aqueous phase containing virus was frozen at –70 ^{0}C. A titer of 4 X 10^7 PFU/mL was obtained.

Plaque assays of HAV and CVB3 were performed by inoculation of monolayers of the appropriate host cells in 60 mm dishes with 0.2-0.25 mL of dilutions of the virus. BGM or FRhK-4 cells were prepared 2-3 days prior to inoculation and were just confluent. After adsorption periods of 1 hr for CVB3 or 90 minutes for HAV, 5.0 mL of overlay medium consisting of MEM + 2%FBS + 1% agarose was added. Inoculated monolayers were incubated at 37 ^{o}C 2-3 days for CVB3 assay and 7-9 days for HAV assay. At the end of incubation, agarose was removed, and the monolayer stained with PBS containing formaldehyde and crystal violet as previously described (Cromeans et al. 1987). Alternatively some CVB3 assays were overlaid with MEM without phenol red, and at day 2-3 an additional agarose overlay containing neutral red was added. Plaques were then enumerated within 24 hrs.

Neutralization of CVB3 infectivity by sCAR was evaluated by classic methods. The stock of CVB3 was diluted to contain a concentration of 100 PFU/ 0.1 mL PBS and 0.1 mL volumes were mixed with equal volumes of dilutions of sCAR purified protein containing for a final concentration of 0(control), 0.035, 0.175, or 0.350 ug of sCAR. After incubation at 37 ^{o}C for 1 hr, 0.2mL volumes of the mixed solutions were inoculated in triplicate onto BGM cells and the plaque assay performed as described above with a neutral red overlay.

Adenovirus culture

Ad2 (ATCC # VR-846, Rockville, MD) and Ad41 (ATCC # VR-930) were propagated in A549 and Graham 293 cells (G293), respectively, in Eagles minimal essential medium (MEM) containing 2%FBS. The viruses were partially purified by chloroform extraction and stored frozen at -80 °C until used for experiments. The infectivity titer of Ad2 was estimated by Most Probable Number (MPN) from CPE observation of inoculated cell cultures. Ad2 stock titer was approximately 10^9 infectious units (IU)/ml. The infectivity titer of Ad41 was estimated by MPN using PCR of viral DNA as the basis for detection in G293 cells. This endpoint was used for scoring infectivity due to ambiguous CPE in cell culture. The infectivity titer of Ad41 stock was approximately 5×10^5 IU/ml.

Chlorine-disinfection and UV treatment of virus preparations

Chlorine-disinfected CVB3 was prepared by standard methods. Sodium hypochlorite (6% household bleach) was used to make a 1 mg/L concentration of free chlorine solution in chlorine demand free water and the concentration of free chlorine was measured by DPD method (Standard Methods, APHA). Virus stock was further purified and washed in 0.01 M phosphate buffer by centrifugal ultrafiltration (Centricon 100, Amicon Corp.) and dispersed by serial filtration through Tween 80-treated 0.2 µm and 0.08 µm porosity polycarbonate filters. Batch disinfection experiments used 1.7 x10^3 PFU of CVB3 with 1 mg/L of free chlorine in chlorine demand free buffer and a contact time of 10 minutes. The disinfected stock had no infectivity titer remaining when 0.2 mL aliquots were assayed in duplicate, and the same end-point RT-PCR titer as the virus control was obtained. In experimental analysis with disinfected virus, cultures were inoculated with sample dilutions having the number of PFU determined prior to disinfection of stock virus; the RT-PCR titer inoculated was the same as the RT-PCR titer of the infectious virus.

Adenovirus was treated with free chlorine by the same method. Batch disinfection experiments used seeded (10^6 PFU/mL) 0.01M phosphate buffers (pH = 7.5) at 4 °C (Sobsey et al 1998B), 1 mg/L of free chlorine and contact times up to 100 min. Viruses with and without free chlorine treatment were subject to culture on A549 cells, DNA detection by PCR, and mRNA detection in cell culture by RT-PCR.

Adenovirus was treated with ultraviolet light (UV). A collimated-beam UV apparatus containing two 15–Watt, low-pressure mercury vapor germicidal lamps emitting nearly monochromatic UV radiation at 254 nm was used. The emitted UV light was directed through a circular opening to provide incident radiation to the surface of the test suspension in a 60- by 15 mm cell culture Petri dish. The UV irradiance was measured with a radiometer (Model IL500; International Light, Inc., Newburyport, MA) that had been factory calibrated, traceable to National Institute of Standard and Technology standards, prior to the study. The measured incident irradiance at the surface of the test liquid was corrected for any nonhomogeneity of irradiation across the surface area of the Petri dish to provide the value for average incident irradiance. Approximately 5×10^5 IU of Ad2 and 10^3 IU of Ad41 were suspended in 5 mL PBS, and exposed to 254 nm UV. The average irradiance in the mixed suspension was determined mathematically by Beer Lambert law over the sample depth, accounting for UV absorbance and incident average irradiance. The exposed UV doses were 0, 50, 100 and 250 mJ/cm^2, which were calculated from the products of exposure time and UV irradiance. Viruses with and without UV exposure were subject to culture in A549 or G293 cells, viral DNA detection by PCR, and mRNA detection in inoculated cell cultures by RT-PCR. Inoculated cell cultures were examined for production of cytopathogenic effects (CPE) for 7-10 days, and mRNA was extracted from cell cultures after 7 days.

WILD-TYPE VIRUS SOURCES, PREPARATION, AND CHARACTERIZATION

Hepatitis A viruses

Three wild-type HAV preparations from stools were used to evaluate replication potential in several cell lines. HLD-2 HAV was obtained from a marmoset 8 days after inoculation with a human virus isolate. An 18% (w/v) stool suspension was prepared in PBS as previously described and has been stored at –70^{0}C for more than 10 years (McCaustland et al., 1982). This preparation has been evaluated for primate infectivity in Tamarins by inoculation of 10-fold dilutions of the HLD-2 stool preparation. One animal infectious dose was 10^{-6}/mL (CDC, unpublished). In preparation for cell culture infection, the 18% preparation was extracted with an equal volume of chloroform. The Q#1 stool preparation was obtained from an individual 2 weeks after onset of hepatitis A symptoms. This individual was infected in a small outbreak associated with a private well (DeSerres et.al. 1999). A 20 % stool suspension was prepared in Dulbecco's PBS by vortexing with glass beads for 2 min followed by centrifugation at 1000 x *g* to obtain the supernatant virus preparation. The virus preparation was extracted with an equal volume of trichlorotriflouroethane followed by extraction with an equal volume of chloroform. Both HLD-2 and Q#1 are genotype 1A HAV, the most common type found in the U.S. and worldwide. A third strain, GA 76, a genotype III HAV, was obtained from a chimpanzee 9 days after inoculation with a human isolate (Khanna et al. 1992). To determine RT-PCR units contained in a given volume of the virus preparations, 10-fold dilutions of the preparations were analyzed. RNA was extracted from dilutions of these preparations with the QIAamp Viral RNA extraction kit (Qiagen, Valencia, CA), and end point titers determined by RT-PCR for positive-strand viral RNA (described below). Titers of HLD-2 and Q#1 were obtained by RT-PCR followed by second round PCR, however the GA76 titer was determined only by first round; the primers used for second round were not specific for GA76. The second PCR performed with an aliquot of the product from RT-PCR yields an increased titer of virus in RT-PCR units.

Four different cell lines were evaluated for replication of wild-type HAV: FRhK-4 and MA-104 (Rhesus monkey kidney cell lines), Caco-2, a human colorectal adenocarcinoma cell line, and PLC-PRF-5, a human hepatoma cell line. T25 cm^2 flasks containing ~2 x 10^6 cells were washed with DPBS and inoculated with 100 to 150 ul of the indicated stool preparations. After incubation at 37 ^{0}C for 90 min, the inoculum was removed, and the cells were washed with DPBS in initial inoculations. Inoculated cells were incubated at 37^{0}C with EMEM + 2% FBS, and maintenance medium was replaced weekly. At the indicated times PI, cells were prepared by two methods. After removal of the maintenance medium, one-half of the monolayer was scraped into 2.0 mL of DPBS. The other half of the monolayer was removed with lysis buffer for extraction of RNA and RF analysis.

Enteroviruses

Enteroviruses were isolated from raw sewage collected after grit removal at an Atlanta wastewater treatment plant. Three liters of raw sewage were collected and a polyethylene glycol precipitation was performed with 0.3 M NaCl and 8% PEG (w/v), 8000 MW as described above. Pellets were resuspended in Dulbecco's PBS, and the concentrate was extracted with an equal volume of chloroform. This process yielded ~50 mL of concentrate for isolation of enteroviruses. Sixty mm plates of BGM cells (UNC/ EPA) were inoculated with 0.3 mL of sewage extract, and after 90 minutes, plates were overlaid with 5 mL of EMEM containing 0.5% agarose. Initially 30 plates were inoculated without any additive in the inocula, and 2 plaque isolates were obtained in 4 days. In a subsequent isolation on 55 plates with the addition of 5 µg/mL of trypsin in the inocula, 9 plaques were isolated at 3 to 5 days PI (post infection). Plaques (5-10 mm in diameter) were visualized macroscopically and confirmed microscopically by the appearance of CPE. All of the agarose over a plaque was removed by aspiration with 2.0 ml-capacity pipette tips and put into 2.0 mL of DPBS. After sonication, the material was chloroform extracted, the interface re-extracted with DPBS, and the upper aqueous layers were combined for analysis. Inoculation at 1:100 dilution of this material into T25 cm^2 flask yielded CPE complete in 4-6 days for all eleven isolates. Amplification of RNA extracted from all the original plaque isolates yielded positives with the panenterovirus primers described below.

Two of the isolates with higher infectivity titers were analyzed in the CDC enterovirus laboratory of Dr. Steve Oberste and identified by sequence analysis. Echovirus 1 and CVB4 were identified based on nucleotide sequence of 321 nt in the VP1 region of the genome (Oberste et al. 1999A and B, and 2000). Replication kinetics of these two viruses were analyzed. Infectivity titers of the original plaque aspirate extracted from the agarose plug were determined by standard plaque assay on BGM cells. Echovirus 1 had a titer of 3.8x10^6 PFU/mL and CVB4 had a titer of 2.4 x 10^5 PFU/ml. These stocks were used in all evaluations of replication kinetics of the wild-type viruses. These enteroviruses were also used for spiking water concentrates and evaluation of recovery with the receptor capture method and for detection by real-time RT-PCR.

PRODUCTION AND PURIFICATION OF THE sCAR VIRUS RECEPTOR

Production, purification and evaluation of sCAR

Tn5-B14 cells (High Five™ cells) were cultured with Ex-Cell 400 media. The sCAR stock was made by inoculation of a T150 of 30-50% confluent Tn5-B14 cells with pass 1 stock of BAC virus (pAcsCAR) at a MOI of 10. After 5 days of incubation, 50 ul of Protease Inhibitor Cocktail for Mammalian Cell Extracts (#P8340, Sigma, St.Louis, MO) was added to 30 mL of harvest from each T150 cm^2flask of cells and the contents frozen and thawed 3 times. After a 30 min spin at 575 x g at 4°C, the supernatant was taken and loaded on columns for purification.

To prepare more BAC virus stock containing the pAcsCAR clone, the BacPAK6 DNA (*Bsu*36 I digest, CloneTech 6144-1) was used for the cotransfection of insect cell Sf9 with pAcsCar, the recombinant plasmid including the gene of the soluble part of CAR. For the cotransfection, cell monolayers in the 1.5mL Grace's Basic Medium were incubated at 27°C for 5 hours after adding the transfection solution including 5µl of 100ng/µl pAcsCar, 5µl of BacPAK6 DNA, 4µl of Bacfectin and 86µl sterile H_2O. Virus was produced by adding 1.5 mL TNM-FM with 10% fetal bovine serum followed by 72 hours incubation.

The sCAR clone was designed to contain amino acids 1 to 236 of the human CAR extracellular domain, upstream of a linker sequence and a FLAG tag consisting of the amino acids DYKDDDDK for detection and purification purposes (Roelvink et al. 1998). The sCAR protein was purified on a 5 mL affinity column with a coupled anti-Flag M2 antibody (Sigma Chemical Co., St.Louis, MO), which was stabilized with TBS (50mM Tris, 150 mM NaCl, pH 7.4). The column was washed alternating with one column volume, 5 ml, of 0.1 M glycine HCl pH 3.5 and then a column volume of TBS. The extract containing sCAR was adjusted to 0.15 M NaCl, loaded and washed with 7 column volumes of TBS. The column was eluted with 6 mL of the 0.1 M glycine-HCl and collected in 1 mL aliquots into tubes containing 25 ul of 1M Tris pH 8.0. The column was washed with TBS for preservation until the next use. The column was used up to 20 times.

The purified sCAR protein was loaded on an SDS-PAGE gel, 20 ul per well, and run in reducing conditions with 12% Tris-Glycine (Novex) 130 V per 2 hr. Proteins were transferred to a 0.2 um PVDF membrane with a Biorad transfer module for 1 hr at 100 V. Colloidal blue staining of the membrane identified protein bands. For specific detection of sCAR on membranes, membranes were reacted with anti-FLAG M2 (Sigma Chemical Co.) at a 1:1000 dilution followed by goat anti-mouse IgG conjugated with HRP from a blotting kit (Biorad, #170-6464) at 33 ul/100ml. The membrane was washed with 0.05% tween20/TBS pH 7.5 between reactants. The FLAG-BAP protein (Sigma Chemical Co.) served as positive control.

Preparation of amine terminated beads, sCAR conjugation and virus capture

Amine Terminated Support magnetic beads, BioSpheres™ were obtained from Biosource (#BNN5001, BioSource International, Camarillo, CA). The stock concentration of 1-micron diameter beads is 50 mg/mL in PBS pH7.5 with EDTA and sodium azide. Bead concentration is 5 x 10^8 beads/mg (2.5 x 10^7 beads / µl stock solution). The beads are composed of superparamagnetic silanized iron oxide and are pre-coated to provide available amine groups for the covalent coupling of proteins or other ligands; each mg of biospheres has 240 nM of available amine groups with a coupling efficiency of 80% for most proteins. The glutaraldehyde-mediated coupling method of Weston and Avrameas (1971).

In order to prepare beads for 7-10 reactions, 20 ul of stock solution beads (5 x 10^8 beads) were resuspended in coupling buffer (0.01 M pyridine) three times, then resuspended and rotated in 5% glutaraldehyde (5mL glutaraldehyde 25% solution plus 20 mL coupling buffer) for 3 hrs at room temperature. Beads were then washed in coupling buffer three times, and 2.5 µg of sCAR in coupling buffer was added to the glutaraldehyde activated particles. After 16-24 hrs of rotation at room temperature, quenching solution (1 M glycine pH 8.0) was added for 30 min and then removed. Beads were washed with washing buffer (0.01 M Tris, 0.1% sodium azide, 0.1% BSA, 0.15 M NaCl, 0.001 M EDTA) three times before rotation for 2-3 hr at RT with BSA blocking solution.

After washing 3 times with PBS to prepare for virus capture, coated beads were resuspended in the appropriate virus dilution and incubated 4-12 hrs with rotation. To prepare for RT-PCR, beads were washed 3X in PBS and resuspended in 140 µl of QIAamp lysis buffer (QIAamp viral kit, Qiagen). After 15 min of incubation, beads were separated magnetically and lysis buffer was transferred to another tube to which 560 µl of EtOH was added. This was applied to the minicolumn and washed per kit instructions. Nucleic acid was eluted from the column with buffer provided and 10µl of the 50 µl eluate was tested in RT-PCR.

RNA EXTRACTION METHODS

Extraction of RNA from water concentrates

An RNA extraction method was developed for the detection of viruses using HAV and CVB3 as models in mock water sample concentrates as described in the results section (3. Real-time RT-PCR and PCR detection of Viruses). Real-time RT-PCR evaluation was used to test the recovery method. Mock concentrates were spiked with known quantities of virus to evaluate recovery efficiency of the methods. The method was first evaluated on a smaller scale with mini columns for RNA extraction. The method was also shown to provide good RNA recovery from 4.0 mL samples on midi columns designed for 1.0 mL sample size. This was accomplished with the use of a special buffer developed in this project. The lysis buffer was developed for use with QIAGEN mini or midi column (Qiagen, Valencia, CA). The following composition of lysis buffer was used for evaluation of the mini or midi columns of QIAgen: 120 g of guanidine thiocyanate (Boehringer Mannheim, USA) was added to 100 mL of buffer (TE buffer containing 5 mM Tris pH 8.0, 0.5 mM EDTA pH 8.0) prepared with nuclease free water (Ambion, Austin, TX). After GuSCN was dissolved, 11 mL of 5 M sodium chloride (Ambion) and 11 mL of 3 M sodium acetate pH 5.5 (Ambion) were added. Carrier RNA (Poly A from Sigma, St Louis) was added to a final concentration of 20 µg/ml. This buffer was evaluated for extraction of water sample concentrates spiked with known concentrations of laboratory viruses including HAV, CVB3, and adenoviruses to establish recovery of RNA and DNA viruses. The buffer was used for all nucleic acid extractions from the concentrated field samples. The following figure illustrates the sequence of reactions for extraction from 4.0 mL of water sample concentrate.

Add equal volume of GuSCN lysis buffer to 4.0 mL water concentrate

↓

After 15 min incubation, centrifuge at 2300 x g for 5 min and discard pelleted debris

↓

Add 8.0 mL of 100% absolute EtOH

↓

Filter through QIAGEN midi column (RNeasy kit) with vacuum (QIAVac) or centrifugation

↓

Wash with 4 mL of 75% EtOH with centrifugation (6300 x g for 5 min), repeat to remove residual EtOH, third spin briefly to remove all EtOH

↓

Elute with 250 µl 10 mM Tris-1mM EDTA (pH 8.0), centrifuge at 2300 x g for 3 min

Figure 2.1. Extraction of RNA from source water concentrates

To test the water concentrates for three viruses, two midi column preparations were used to prepare RNA. The volume of RNA sample obtained from elution of each column was about 200 µl. The approximately 400 µl total sample was divided into 3 fractions to separately test for each of three virus types. The RNA from the column was reprecipitated with $1/10^{th}$ volume of 5 M ammonium acetate, 2.5 volumes of EtOH and 1 ul of glycogen (20mg/mL stock per 60 µl of RNA. After vortexing briefly, incubating at –20 ^{o}C for 30-60 min, followed by centrifuging at 14000 rpm for 15 min 4° C, the pellet was washed once with 500 µl of 75% cold ethanol. Pellets were air dried and suspended in 5 µl nuclease free water. These precipitates were either tested immediately or stored at –70 ^{o}C until assayed by PCR or RT-PCR for enteroviruses, adenoviruses, or NLV.

Drinking water sample concentrates of one mL were obtained from 100 L of water as described above in water concentration methods with the use of filtration on positively charged filters at the utility. Filters shipped to UNC were eluted with amino acids and the eluate concentrated with PEG. Of the one mL sample concentrate, 300 µl was lysed with the GuSCN lysis buffer and nucleic acid separated on the minicolumn to enteroviruses, adenoviruses, and Noroviruses.

Preparation of viral RNA for Positive Strand and RF assays of HAV and enterovirus

The choice of method depended upon the source of RNA (viral preparations in aqueous samples or infected cells) and type of RNA needed for analysis. Positive-strand viral RNA was extracted differently than the double-strand RF RNA (containing the positive strand and complimentary negative strand). RF RNA was extracted directly from the infected cells with appropriate lysis buffers. The positive strand viral RNA was extracted from a stock virus preparation, from infected cells prepared as an aqueous preparation, or from supernatant culture medium of infected cells. Figure 2.2 shows the general scheme for preparation of the different types of RNA and the related assays that were performed.

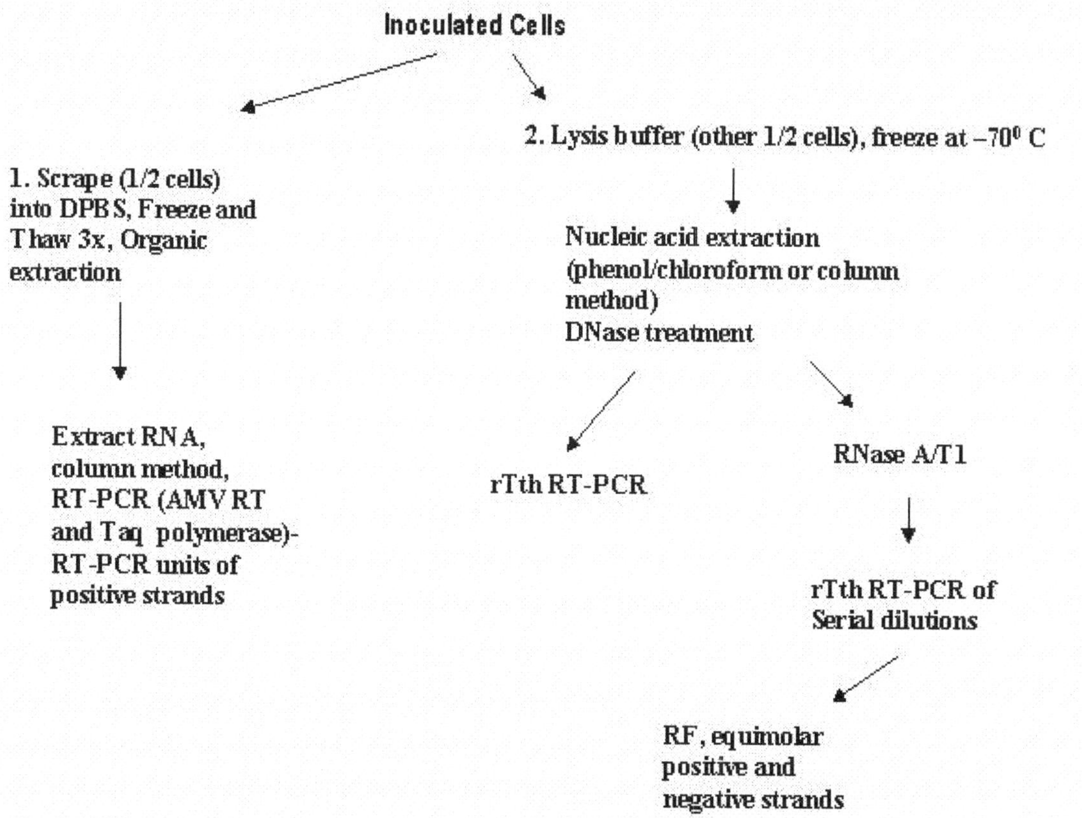

Figure 2.2. Preparation of RNA by different methods from infected cells

Viral RNA was extracted from dilutions of HAV or CVB3 stock virus with the QIAamp Viral RNA extraction kit (Qiagen, Valencia, CA). The sample was lysed under denaturing conditions, adjusted to binding conditions and applied to spin columns, where RNA was bound and contaminants washed away. RNA was eluted from the column with water or 0.1 mM EDTA. This extraction method was also used for detection of total RT-PCR viral units in infected cells harvested in DPBS. Infected cells were harvested in DPBS, frozen and thawed three times, and chloroform extracted. RNA was extracted from 10-fold dilutions of the cellular virus preparations with the QIAamp viral column method.

To extract RNA from infected cells for analysis of RF production, the Totally RNA (Ambion, Austin, TX) kit was used with instructions provided. Two and one-half mL of denaturation solution was added to ~ 1 x 10^7 cells at the indicated times PI. This solution was extracted with phenol:chloroform:isoamylalcohol, the aqueous extract was re-extracted with acid-phenol, and the final aqueous extract precipitated with isopropanol 2 hr or overnight at –20 °C. The RNA pellet obtained by centrifugation at 10,000 x *g* was washed with 70% ethanol, and then DNase treated.

RQ1 RNase-free DNase (Promega, Madison, WI) was used to degrade any DNA remaining in the RNA preparation. The DNase digestion was performed as described using the

reaction buffer provided. Forty units of DNase was used for ~1 x 10 ^{7} cells or ~40 μg of RNA. After incubation at 37 °C for 30 min, the RNA was treated with 20 mg/mL proteinase K and 0.5% SDS for 15 min at 37 °C, followed by re-extraction with phenol:chloroform:isoamylalcohol and by isopropanol precipitation. An aliquot of this DNase-treated RNA sample was reserved for analysis, and the remaining sample was treated with RNase at a 1:2000 dilution of RNase cocktail of A and T1 (Ambion, Austin, TX) containing RNase A at 500 U/mL and RNase T1 at 20,000 U/mL. RNase cocktail digestion was performed in a buffer containing 300 mM NaCl, 10 mM Tris pH 7.5, and 5 mM EDTA. Reactions were incubated for 30 min at 30 °C, treated with 20 mg/mL proteinase K and 0.5% SDS for 15 min at 37 °C, and extracted with phenol:chloroform:isoamylalcohol, followed by isopropanol precipitation with the addition of 40 μg glycogen as carrier. Pellets containing the double-stranded RNA were resuspended in DNase- and RNase-free water or DNase- and RNase-free water + 0.1 mM EDTA. Preparations were either used immediately in RT-PCR reactions or frozen at
–70°C for future use when 0.1 mM EDTA was included.

Nucleic acid purification for Adenovirus mRNA studies

Messenger RNA was purified from cultured cells by the following procedure. Either A549 or G293 cells were cultured in T25 flasks for 4-5 days at 37 °C. Cell monolayers were inoculated with the pre-determined number of Ads in one mL of PBS, and incubated with periodic gentle mixing for 1 hour at 37 °C. After the inoculum was removed, 10 mL of MEM containing 2% fetal bovine serum was added to the cell monolayer and cultured at 37 °C. After the indicated incubation period for virus replication, the cell monolayer was washed with 5 mL of PBS, and then treated with 5 mL of 0.25% trypsin (Sigma) in PBS for 5 min. Cells were transferred to a polypropylene centrifuge tube, harvested by centrifugation at 300 x g for 5 min., and the supernatant was discarded. The recovered cell pellet was lysed by 3% β-mercaptoethanol and guanidium isothiocyanate (GITC), and then homogenized by QIAshredder (Qiagen, Valencia, CA).

The mRNA in cell lysates was purified by oligo-dT latex following the standard protocol of the Oligotex direct mRNA kit (Qiagen, Valencia, CA) to a final volume of 60 μl. To remove any contaminant DNA, 10 μl of mRNA was incubated for 10-15 minutes with 2 μl of RQ1 RNase-free DNase, 2 μl of DNase buffer (Promega, Madison, WI), and 1 μl of RNase inhibitor (Promega) in a 20 μl volume. Residual DNase was inactivated by DNase stopping solution (Promega). A PCR reaction was performed to check for any residual contaminating DNA. The absence of DNA was confirmed by electrophoresis on a 2% agarose gel followed by ethidium bromide staining. Viral DNA was extracted using the QIAamp viral kit following the standard protocol (Qiagen, Valencia, CA).

RT-PCR AND PCR

HAV and enterovirus detection of positive viral strands

Oligonucleotide primers from the HAV genomic capsid region previously described for RT-PCR of HAV were used (Margolis and Nainan 1990). The 2167-2192 forward primer (2167-2192 relative to the genome sequence of HM-175 by Cohen and colleagues in 1987) and reverse primer (2390-2414 relative to the HM-175 sequence) yield an amplicon of 248 bp when visualized on 2% agarose gels with ethidium bromide staining. RT-PCR was performed similar to the previously described methods with some modifications (Cromeans et al. 1997). Twenty-five µl of RNA was included in the RT reaction with AMV RT (Roche, Mannheim, Germany) for a final volume of 50 µl containing 5.0 mM $MgCl_2$, 200 ng of reverse primer, and 0.2 mM dNTP. After 60 min at 42 °C, Taq DNA polymerase (Roche) and the forward primer were added to generate a final 100-ul reaction containing 2.5 mM $MgCl_2$. After 95 °C for 2 minutes, 40 cycles of 95 °C for 30 sec, 55 °C for 1 min, and 72 °C for 75 sec were performed followed by 7 min extension at 72 °C. A second round of PCR was performed with primers amplifying nucleotides 2195 to 2380 relative to the genome sequence of HM-175 of Cohen et al. (1987), yielding an amplicon of 185 bp. The conditions of the second round PCR were the same as the first round, and 30 cycles of the same cycling conditions were used.

Enterovirus amplification of RNA was performed with primers previously described for the 5' NTR region of the genome (Schwab et al. 1995). All enterovirus experiments were performed with these primers. The same conditions described for the HAV RT-PCR were used. Thermocycling conditions were slightly different. After 95 °C for 2 min, 40 cycles of 95 °C for 45 °C sec, 55 °C for 45 sec, and 70 °C for 45 sec were performed, followed by 7 min extension at 70 °C. CDC primers generate an amplicon size of 114 base pairs with primers
5' –ACACGGACACCCAAAGTAGTCGGTTCC -3' and
5'- TCCGGCCCCTGAATGCGGCTAATCC – 3'.

Immunocapture RT-PCR methods

The HAV immunocapture RT-PCR was performed similar to the published method (Cromeans et al. 1997) with some modifications. Microfuge tubes (0.65 ml) were coated overnight at room temperature with 0.65 ul of monoclonal antibody K2-4F2 (GL Limited, Melbourne, Australia) diluted in PBS. Experiments have been performed to determine that K2-4F2 is equally sensitive to the hyperimmune anti-HAV rabbit IgG used previously in the capture of HAV. The end-point of detection in a 10 fold dilution series is the same as end-point detection with hyperimmune anti-HAV rabbit IgG preparation in three trials. Tubes were washed with PBS + 0.5%BSA (PBS/BSA) 2x before addition of virus dilutions in PBS, incubated 2 hr at 42 °C, washed 2x with PBS/BSA, and the RT buffer mix added in 43.5 ul. The tubes were heated at 95 °C and frozen in a dry ice methanol bath. After thawing, the RT mix of 6.5 ul was added and the 50 µl reaction incubated 1 hr at 42 °C, then inactivated at 95°C, kept on ice for the addition of 50 ul of mix containing Taq polymerase. The forward primer and additional buffer were added and kept on ice until placed in the thermocycler heated to 60°C.

The immunomagnetic bead system was used to develop an antibody-mediated capture system with CVB3 for comparison with the sCAR protein capture system. Biosource beads were coated with 1.5 µg MAB to CVB (Chemicon International, INC, Temecula, CA) per 5 x 10^7 beads with the same methods as used for coating with sCAR protein. After capture and washes with PBS, the separated beads were treated by heat release at 94 °C for 2 min and 10 µl was tested in RT-PCR.

RF Detection

The EZ rTth RNA PCR kit (Perkin Elmer, Branchburg, NJ) was used to evaluate RNA extracted from infected cells. Briefly the components supplied, including 5X buffer, rTth DNA polymerase and 25 mM $Mn(OAc)_2$ were used; dNTPs (Roche, Indianapolis, IN) were used at a final concentration of 600 μM. Ten microliters of RNA were included in each reaction; controls included samples with no RNA. Control reactions with DNase- and RNase-treated RNA and no added primers were also performed. For positive strand detection, the reverse primer was included in the RT reaction at 200 ng/reaction, and for negative strand detection, the forward sense primer was included in the RT reaction. Forward and reverse reactions were performed on the same dilutions/samples for comparison and determination of the RF. Reverse transcription conditions were 60 °C for 30 min, followed by 95 °C for 2 min. PCR was performed with 40 cycles of: 94 °C for 1 min, 60 °C for 1 min, 72 °C for 1min, followed by 7 min extension at 72 °C. Reaction products were examined on 2% agarose gels containing 5μg/mL of ethidium bromide. Bands were visualized and photographed on a Stratagene Eagle Eye II system (La Jolla, CA).

Adenovirus PCR and mRNA RT-PCR methods

The genomic sequences of adenoviruses were obtained from GenBank, National Center for Biotechnology Information (NCBI). The primers (AdC-E1AF/AdC-E1AR, AdF-E1AF/AdF-E1AR) complementary to E1A in the serotype-specific region (Ads group C; 2/5 and F 40/41) were designed from alignments prepared by the program BioEdit (version 5.0.9) with resulting amplicons being 260 bp and 280 bp for Ad2 and Ad41, respectively. In addition to serotype-specific primers, Ads specific primers (Hex1/Hex2) to detect hexon gene sequences conserved among all recognized human Ads serotypes were used, yielding an amplicon of 482 bp (Xu et al. 2000). Amplified DNA by Ad group specific primers (Hex1/Hex2) was subject to hemi-nested PCR (Hex1/Hex3), with the resulting amplicon being 443 bp, when higher detection sensitivity was required. The oligonucleotide primers used in the mRNA study are summarized in Table 2.1. The sensitivities of primers were evaluated with viral genomic DNA heat-released from serially diluted virus stock solutions at 95 °C for 2 min.

Table 2.1
Oligonucleotide primers for nucleic acid amplification of Ad

Ad Species	Primer	Target gene	PCR	Sequence (5′→3′)	Amplicon size (bp)
A-F	Hex1	Hexon	First	TTCCCCATGGCICA(CT)AACAC	482 (18)
	Hex2		First	CCCTGGTA(GT)CC(AG)AT(AG)TTGTA	
	Hex1		Nested	TTCCCCATGGCICA(CT)AACAC	443
	Hex3		Nested	AGGAACCA(AG)TC(CT)TT(AG)GTCAT	
2/5	AdC-E1AF	E1A	First	CCACCTACCCTTCACGAACT	260
	AdC-E1AR		First	CTCGTGGCAGGTAAGATCG	
40/41	AdF-E1AF	E1A	First	GGGAACTGGGATGACAT	280
	AdF-E1AR		First	CCSTCTTCATAGCATTTC	

PCR was performed in 50-μl volumes containing 45 μl of reaction mixture (10 mM Tris-HCl [pH 9.0], 50 mM KCl, 0.1% Triton X-100 [Promega, Madison, WI], 1.5 mM $MgCl_2$, 200 μM each deoxynucleoside triphosphate, 0.6 μM each primer, 2.5 U of *Taq* DNA polymerase [Invitrogen, Carlsbad, CA]) per reaction and 5 μl of sample DNA extract. Reverse transcription (RT) was performed on 5 μl of RNA extract and 4 μl of antisense mix, and heated at 94 °C for 2 min to remove the secondary structure of RNA. The RT reaction was performed in 15 μl volumes using 10X PCR buffer (pH = 8.3), containing 50 pM of primer, 3mM $MgCl_2$, 1mM nucleotide mix, 5 U of AMV-RT (Promega), and 20 U of RNase inhibitor (Promega) at 42 °C for 1 hour, followed by inactivation of RTase at 94 °C for 2 min. PCR was performed in 50 μl volumes containing 45 μl of reaction mixture (10X PCR buffer [Promega]: 10 mM Tris-HCl [pH 9.0], 1.5 mM $MgCl_2$, 50 mM KCl, 200 μM each nucleotide, 30 pM each primer, 2.5 U of Taq DNA polymerase [Invitrogen]) per reaction and 5 μl of RT reaction. The amplification reaction was carried out in a thermocycler (model#: PTC-200; MJ Research, Watertown, MA) with preliminary denaturation for 5 min at 94 °C, followed by 40 cycles of denaturation at 94 °C for 1 min., annealing at 45 °C (E1A primers) or 50 °C (hexon primers) for 1 min, and primer extension at 72 °C for 2 min., and final extension at 72 °C for 5 min. In all mRNA RT-PCR reactions, RT-PCR without RTase also was performed in order to prove no DNA contamination in mRNA extracts. Hemi-nested PCR was performed in 50 μl volumes containing 49 μl of reaction mixture (10X PCR buffer [Promega]: 10 mM Tris-HCl [pH 9.0], 1.5 mM $MgCl_2$, 50 mM KCl, 200 μM each nucleotide, 0.6 μM each primer, 2.5 U of Taq DNA polymerase [Invitrogen]) per reaction and 1 μl of either undiluted or 10-fold diluted RT-PCR reaction. The amplification reaction was carried out in a thermocycler with preliminary denaturation for 5 min at 94 °C, followed by 40 cycles of denaturation at 94 °C for 1 min., annealing at 50 °C for 1 min, and primer extension at 72 °C for 2 min. and then final extension at 72 °C for 5 min. Ten μl of each reaction product was separated (1 hour, 120V) on 2% agarose gels, stained by ethidium bromide, and visualized by UV transillumination.

REAL-TIME RT-PCR AND PCR

Development of Taqman probes for detection of HAV and enteroviruses

The nucleotide sequence of primers and fluorogenic probes used in amplification and detection of HAV and enterovirus are presented in the Table 2.2. The reporter dye FAM (6-carboxyfluorescein) or ROX was conjugated at the 5'ends of these probes, and the quencher dye Black hole quencher (BHQ) 1 and Black hole quencher (BHQ) 2 were conjugated in the respective probes at the 3' end. The ROX-labeled probe was used for detecting the 89-bp RNA-dependant RNA polymerase gene fragment of HAV, and the FAM-labeled probe facilitated the detection of the 145-bp 5' nontranslated region fragment of Coxsackievirus B3. Another FAM-labeled probe for detection of 89-bp 5' nontranslated region fragment of HAV was prepared for use in LightCycler instrument for non-multiplex RT-PCR and for quantification studies. The Primers and probes were synthesized at the core facility at CDC-Atlanta.

Table 2.2

Oligonucleotide primer/probe sequences for HAV and EV Real-time RT-PCR

Primer or Probe	Primer or Probe	Sequence (5'-3')	Tm (°C)	Position
HAV				
Forward	HAV-F	5'GGTAGGCTACGGGTGAAAC -'3	59.4	392-410
Reverse	HAV-R	5'GGTAGGCTACGGGTGAAAC -'3	59.8	480-461
Probe	HAVRox	5'CTTAGGCTAATACTTCTATGAAGAGATGC-3	61.4	413-441
Enterovirus				
Forward	EV-R	5' CCTGAATGCGGCTAATCC –3'	61.5	455-472
Reverse	EV-R	5' CCTGAATGCGGCTAATCC –3'	61.5	599-580
Probe	EVFam	5 CCGACTACTTTGGGTGTCCGT –3'	65.6	539-559

GenBank accession numbers for nucleotide position of specific primers and probes - Sequence M19786 used as reference for the nucleotide position of HM175; NC_001473.1 for coxsackievirus B3 (CVB3) serves as reference sequence.

Tm, melting temperature, was calculated based on nearest-neighbor method.

RNA extraction with mini and midi columns

Ten-fold serial dilutions were made of HAV or CVB3 stock virus preparations in DPBS. Ten ul of sample was diluted to 100 µl with PBS and added with 100 µl of GuSCN buffer (the laboratory developed buffer described above) and incubated for 10 min. Absolute ethanol (200 µl) was added to the mixture, vortexed and applied to a QIAgen spin column, centrifuged (8,000 x g for 1min) and washed once with 75% absolute ethanol (centrifuge 20,000 x g for 90 s) followed by a dry spin. RNA was eluted with 50 µl of 10 mM Tris-1mM EDTA (pH 8.0), centrifuged at 8000 x g for 1 min and stored at –20 °C until use.

In the second method for comparison with the developed lysis buffer, RNA extractions were purified from stock virus preparations of HAV and CVB3 using QIAamp viral RNA kit (Qiagen, Valencia, CA) per the kit instructions. Briefly tenfold serial dilutions were made of HAV or CVB3 tissue culture supernatant in PBS. Ten ul of sample diluted to 100 ul with PBS and added with 400 ul of AVL and incubated for 10 min. Absolute ethanol (400 µl) was added, vortexed and applied to the spin column, centrifuged, and washed twice with buffers AW1 and AW2. RNA was eluted with 50 ul of 10 mM Tris-1mM EDTA (pH 8.0) and stored at –20 °C until use.

For extraction of RNA on the midi columns (obtained from the RNeasy kit of Qiagen or from Omnitek {Norcross, GA}), the protocol shown in Figure 2.1 was followed. Briefly, an equal volume of the newly developed GuSCN buffer (described above under Extraction of RNA from water concentrates) was added to the sample, incubated, and after addition of an equal volume of 100% EtOH, added to the midi column. Either centrifugation or vacuum was used to filter through the column, followed by two 75% EtOH washes. Finally the column was eluted with 200 µul 10 mM Tris-1mM EDTA (pH 8.0). In some experiments this was followed with further RNA concentration by the additions of 1/10th volume of 5 M ammonium acetate, 2.5 volumes of EtOH and 1 ul of glycogen (20 mg/ml) per 60 ul of RNA sample.

RT-PCR and Duplex RT-PCR

A 20 μl reaction contained RT-Taq DNA polymerase, QuantiTect ^{Tm} Probe RT-PCR (Qiagen, Valencia, CA) buffer, dNTPs, 0.5 uM primers, 0.25 uM of TaqMan probe (in the multiplex 250 nM of each primer, 125 nM of each TaqMan probe) and 2 ul of RNA from stock virus preparations or environmental isolates. QuantiTect^{Tm} SYBR Green RT-PCR (Qiagen, Valencia, CA) was used in the reactions when the TaqMan probe was not included. After 30 min of 50 °C and denaturation at 95 °C for 15 min, 50 cycles of 95 °C for 10 s, 55 °C for 20 s (45 s multiplex) and 72 °C for 15 s, signal was acquired at every cycle at the end of annealing at 55 °C for HAV and/or CVB3.

The QuantiTect™ Probe RT-PCR 2X buffer containing RT, Taq DNA polymerase, nucleotides, and $MgCl_2$. was used for the duplex reactions. To each reaction were added 2.5 μl of template RNA and volumes of primer to yield a final concentration of 250 nM each of HAVF-HAVR primer and 250 nM each EVF-EVR primer. The fluorogenic probe was added to a final concentration of 125 nM each. Amplification and detection was performed with the SmartCycler (Cepheid, Emeryville, CA) under the following conditions: (1) RT step for 30 min at 50 °C, (2) 15 min at 95 °C to activate Taq Polymerase, and (3) 45 cycles of 15 s at 95 °C, 20 s at 55 °C and 15 s at 72 °C, with monitoring of fluorescence during the annealing phase. During amplification the detector acquired fluorescence in channel 1 (505-537 nm) for the FAM probe and channel 4 (605-800 nm) for the ROX probe in a 25 μl reaction tube. Samples containing both viruses were prepared by the addition of 10 μl aliquots of 10-fold serial dilutions of each virus to 10 mL of PEG water sample concentrate from either river or lake.

For the evaluation of potential interaction of the different fluorophores, a non-multiplex RT-PCR was also performed in two independent RT-PCR reactions using the same master mix as was used in multiplex RT-PCR. Each non-multiplex reaction contained the multiplex primer pair with a FAM labeled probe and a ROX labeled probe. The FAM signal was read in the LightCycler only for optimization and to show sensitivity.

Adenovirus Real-time PCR

The TaqMan I forward and reverse primers shown in Table 2.3 were designed from the hexon region to amplify a 135 bp fragment of adenoviruses of species C, D and F, therefore detecting Ad 2 and 5 (species C) and Ad 40 and 41 (species F). The TaqMan II primers amplify a 118 bp fragment to specifically detect species F (Ad 40 and 41).

Table 2.3
Oligonucleotide primer/probe sequences for Adenovirus Real-Time PCR

Primer or Probe	Name/Position (5'-3')	Sequence (5'-3')	T_m (°C)
TaqMan I[a]			
Forward primer	Ad-GF (18891-19910)	5' GGACGCCTCGGAGTACCTGA -'3	59.4
Reverse Primer	Ad-GR (19025-19007)	5' CGCTGIGACCIGTCTGTGG –'3	59.8
Probe	AdG FAM (18939-18960)	5' CACCGATACGTACTTCAGCCTGGG-3'	61.4
TaqMan II[b]			
Forward primer	Ad-FF[b]	5' AACTTTCTCTCTTAATAGACGCC–3'	61.0
Reverse Primer	Ad-FR	5' AGGGGGCTAGAAAACAAAA -3'	61.2
Probe	AdF FAM	5' CTGACACGGGCACTCTTCGC –3'	70.1

[a]Position based on GenBank accession # J01971.1 hexon region of adenovirus 2.
[b]The relative orientation of forward and reverse primers positioned at 619-640 and 736-723 respectively. The TaqMan probe was labeled with FAM at its 3' end and black hole quencher at 5'(692-673). Position based on GenBank accession # X16583.1 fiber protein gene of adenovirus 41. Tm = Melting temperature was calculated based on nearest-neighbor method.

Real-time PCR was performed in LightCycler glass reaction capillaries with 2 µl DNA in 20 µl volumes with 0.25 uM primers, and 150 nM fluorogenic probe. Nucleotides, Taq DNA polymerase, and buffer were included in the QuantiTect[Tm] Probe PCR (Qiagen, Valencia, CA) kit. A typical protocol took approximately 90 min to complete and included hot-start denaturation for 15 min, followed by 50 cycles of a 95 °C denaturation for 10s, 55 °C annealing for 20 s and extension at 72 °C for 15s. The fluorescence was acquired at the end of each annealing step. All amplification reactions were carried out in duplicate. To confirm the specificity of PCR, products from each primer pair were subjected to agarose gel electrophoresis.

Norovirus detection

Primers and the TaqMan probe designed for Genogroup I NLV detection are shown in Table 2.4. The probe (JTMG1P) was labeled 5' with FAM and BHQ on the 3' end the same as other probes described in the study.

Table 2.4
Primers and TaqMan probe for Norovirus Genogroup 1 Detection

Name	Polarity	Location	Primer sequence
JTMG1F	Forward	5282-5299	GCIATGTTCCGITGGATG
JTMG1R	Reverse	5347-5328	GTCCTTAGACGCCATCATCAT
JTMG1P	Forward	5289-5311	TGTGGACAGGaGATCGCaATCTC

Positions correspond to the Norwalk virus genomic RNA, GenBank accession # M87661 (Le Guyader et al.1996).

RNA Extraction protocol for stool suspensions

The NucliSens automatic extractor is a bench-top semi-automatic instrument using the BioMérieux proprietary silica based extraction technology with guanidine thiocyanate (GuSCN) for lysis. The method is modified from the Boom method that uses silica to bind RNA after lysis with a guanidine thiocyanate buffer (Boom et al., 1990). The Nuclisens closed system was designed to eliminate both sample-to-sample carryover and environmental contamination throughout the extraction process. It was used for RNA extraction of some of the Norovirus stool samples. A pea sized stool sample was added to 2 mL of 50% Vertrel XF (Dupont, CT; prepared by mixing with equal amount of nuclease free water). The sample was vortexed thoroughly and centrifuged at 3000 x g for 5 min. The top aqueous phase was processed for RNA extraction. Total nucleic acid was extracted from 100 µl of sample with automated isolation reagents supplied by Nuclisens (Organon Teknika, Boxtel, NL) on the automated nucleic acid extraction instrument, Nuclisens extractor (Nuclisens, Organon Teknika, Boxtel, NL). Briefly, lysis buffer of 900 µl was added to 100-µl samples and incubated for 10 min at room temperature. After 10 min, 50 µl of silica solution was added and the sample vortexed and then transferred to individual cartridges, together with a denaturing solution and silica particles, which were sealed and mounted on the instrument. Subsequent washing and elution buffers were pumped automatically through each cartridge separately and the RNA was eluted to final volume of 50 µl with buffer.

For manual isolation of RNA, the stool samples were diluted to ~10% in PBS and were clarified by vortexing and centrifugation at 3000 x g for 15 min. RNA was extracted from 100 µl of clarified stool samples using QIAamp viral RNA kit (Qiagen, USA) according to the instructions supplied with the kit. The RNA was eluted to a final volume of 50 µl.

TaqMan RT-PCR for Norovirus

TaqMan RT-PCR reactions were carried out in glass capillaries in a final volume of 10 µl. Nucleotides, RT-Taq DNA polymerase, and buffer were included in the LightCycler-RNA Hybridization probe kit reaction mix (Roche Molecular Biochemicals, USA). Template RNA (2.0 µl each) was added to 8 µl of a LightCycler-RNA Hybridization probe kit mixture containing 250 nM each JTMG1F - JTMG1R primer, 250 nM fluorogenic probe (JTMG1P) and a Mg^{2+} concentration of 5mM. The RT-PCR was carried out at 55 °C for 30 min, followed by denaturation at 95 °C for 45s, then followed by 40 cycles with 95 °C denaturation for 5s, 55 °C annealing for 35 s and extension at 72 °C for 15s; fluorescence was acquired at the end of 55 °C at each cycle. Specificity was also confirmed by agarose gel electrophoresis.

Table 2.5 illustrates the selection of the forward primer and mismatches between several virus types within Genogroup I. Table 2.6 describes the forward and reverse primers and probe selected for broad detection of Genogroup I NLV.

Table 2.5
Mismatches between genomes of Norovirus in the Forward primer

Virus type	Forward primer (5'-3') sequence
Desert shield	GCCATGTTCCGCTGGATG
Southampton	GCCATGTTCCGTTGGATG
Norwalk	GCCATGTTCCGCTGGATG
Chiba	GCTATGTTCCGCTGGATG
FW	GC*ATGTTCCG*TGGATG

FW-Nucleotides that differ are depicted as * resulted in the design of
GI Forward primer GCIATGTTCCGITGGATG

Table 2.6
Primers and TaqMan probe for Norovirus G1 Detection

Name	Polarity	Location	Primer sequence
JTMG1F	Forward	5282-5299	GCIATGTTCCGITGGATG
JTMG1R	Reverse	5347-5328	GTCCTTAGACGCCATCATCAT
JTMG1P	Forward	5289-5311	TGTGGACAGGaGATCGCaATCTC

Positions correspond to within the Norwalk virus genomic RNA, GenBank accession # M87661 (Le Guyader et al.1996).

CHAPTER 3
RESULTS AND DISCUSSION

DETECTION OF RF IN HAV AND ENTEROVIRUS IN INFECTED CELLS

Detection of RF in cells infected with laboratory strains of virus

The RF detection method was developed and evaluated initially using a laboratory-adapted strain of HAV HM-175 (clone 24A) in infected FRhK-4 cells. First, cells infected at a high MOI (concurrently infecting all cells in the monolayer) were harvested and evaluated with different concentrations of RNase (data not shown). The RNA extraction and treatment described in methods and materials was optimal for production of the expected equal quantity of positive and negative strands in the RF assay. An excess concentration of RNase removed even the double stranded structures and a low concentration of RNase left single strands intact and therefore yielded detection of positive strands not paired with negative strands. After establishment of the optimum concentration of RNase, experiments were done in which cells were infected at lower MOI's and examined at different harvest times. Laboratory stain virus was investigated to evaluate methodology prior to application to wild-type strains of virus. The results for two different MOI's and harvest times are shown in Table 3.1.

Table 3.1.
HAV RNA detection in cells infected with laboratory strain HAV (HM-175, clone 24A) at varying MOI and days PI

	DNase treated	DNase treated	DNase and RNase treated	DNase and RNase treated
MOI (Days PI)	Positive Strand End-point	Negative Strand End-point	RF (Positive Strand)	RF (Negative Strand)
0.1 (2 da)	10^7	10^3	10^3	10^2
0.1 (4 da)	10^6	10^4	10^5	10^4
0.01 (5 da)	$\geq 10^8$	10^5	10^3	10^3
0.01 (7 da)	10^9	10^5	10^5	10^4

The data in Table 3.1 indicates that at lower MOI's and at different times after infection, the replicative forms can be identified after DNase and RNase treatments. At all MOIs and time points, all cells remained attached, although at 4 day with 0.1 MOI and 7da with 0.01 MOI, CPE was apparent in that 30-70% of cells exhibited rounding although all cells were attached. The end-point titration results of RF (positive and negative strands after DNase and RNase treatment) suggest that there is an increase in quantity of RF between 2 and 4 days PI and 5 and 7 days PI at higher and lower MOIs, respectively. The replication kinetics of laboratory strain virus were used to validate methodology and were not further investigated. Other investigators have found that after appearance, the negative strands are present in the same quantity for weeks after infection with a slowly replicating laboratory strain of HAV (Schmitz and Dotzauer, 1998), although this work was performed with a different methodology for detection of negative strands than the RF method. Our data also suggests that the RF are stable over time (at least days) or that any degradation of RF is followed by synthesis of new RF such that a constant level is maintained.

The method developed for RF detection also was applied to a model enterovirus, CVB3. Initially, BGM cells were infected with an MOI of 2 PFU/cell and harvested at 8 hr PI. Two flasks were infected, one flask was harvested for determination of virus infectivity titer and the other used to detect RF by harvesting with lysis buffer, and preparation of RNA with DNase and RNase treatment as previously described for HAV-infected cells. The infectivity titer of the harvest from the flask of cells infected at high MOI was 2 x 10^7 PFU/ml. Results in Figure 3.1 demonstrate that after DNase and RNase treatment, the quantity of positive and negative strand RNA is equal, since a dilution of 10^3 was the end-point of amplicon detection in the assays for both positive and negative strands after DNase and RNase treatment (lanes 8 and lane 17; lane 8 appears faint here, but it was visible clearly on the gel). At a relatively high MOI, the method successfully detects RF production in CVB3 infected BGM cells.

M 10^{-3} 10^{-4} 10^{-5} 10^{-6} 10^{-7} 10^{-1} 10^{-2} 10^{-3} 10^{-4} 10^{-1} 10^{-2} 10^{-3} 10^{-4} 10^{-5} 10^{-1} 10^{-2} 10^{-3} 10^{-4} M

Result + + + + + + + + - + + + + + + + + -

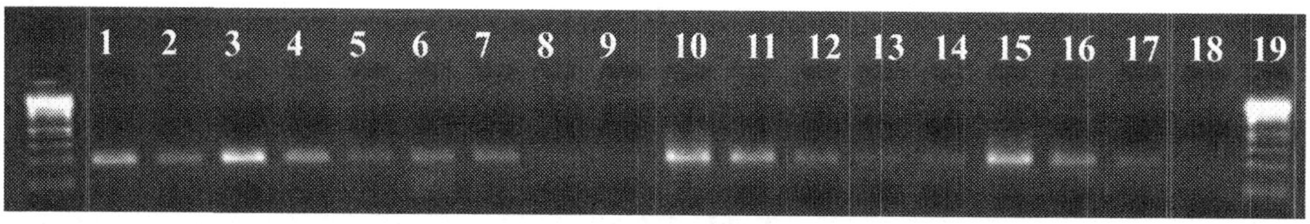

Figure 3.1.Detection of RF in CVB3 infected cells with RT-PCR. Amplicons in lanes 1-9 were generated with primers added for positive strand RNA detection, and amplicons in lanes 10-18 were generated with primers added for negative stand RNA detection. In lanes 1-6 amplicons were produced from RNA dilutions of 10^3 to 10^7 of DNase treated RNA. In lanes 6-9, amplicons were produced from RNA dilutions of 10^1 to 10^4 of RNase A/T1 treated RNA. In lanes 10-14, amplicons were produced from RNA that had been DNase treated and diluted 10^1 to 10^5, and lanes 15-18 amplicons were produced from RNA had had been RNase treated and tested at dilutions of 10^1 to 10^4. The detection sign (+/-) was based on visual examination of the agarose gel for an amplicon of the correct size.

BGM cells infected at a range of lower MOIs (0.01-0.00001 PFU/cell and equivalent to 10^4 -10 PFU/flask) of CVB3 were analyzed for the presence of replicative forms at 24 hr PI. In a separate experiment cells infected at 0.00001 MOI were analyzed at 48 hr PI. Results of these experiments are shown in Table 3.2. As MOI was decreased, decreasing quantities of positive strand RNA (measured after DNase treatment, second column in Table 3.2) were detected in infected cells. As expected, this result reflects a slower rate of viral replication with lower MOI. The quantity of positive strand RNA detectable after DNase and RNase treatment is 3 to 4 orders of magnitude lower after only DNase treatment. This reflects the excess of positive strands present in the infected cells. These genomic single (positive sense) strands are in excess of the negative strand and are not degraded by DNase treatment. After treatment with both enzymes, the quantities of RNA detectable by RF (positive and negative strand) assays were approximately equal, indicating that only double stranded replicative forms were detected. Results in Table 3.2 demonstrate that, as expected, an increased quantity of RF was detected at 48 hr PI as compared to 24 hr PI at the lowest MOI (equivalent to ~20 PFU/flask). Infection with a lower input of virus requires longer incubation time to obtain a quantity of RF present that is the same as that obtained with a higher input of virus in less time. This will be an important consideration in evaluation of infection with an unknown inoculum, when a low quantity of virus is expected in the sample.

Table 3.2.
End-point detection of viral RNA in CVB3 infected cells

MOI Hrs PI	DNase treated	DNase treated	DNase and RNase treated	DNase and RNase treated
	Positive strand	Negative Strand	Positive strand	Replicative form
0.01 24 hr PI	10^8	10^4	10^5	10^4
0.0001 24 hr PI	10^7	10^3	10^3	10^3
0.00001 24 hr PI	10^6	10^3	$\geq 10^1$	10^2
0.00001 48 hr PI	10^7	10^4	10^4	10^4

To determine the time after infection of appearance of RF, 4 PFU of CVB3 was inoculated per T25 cm^2 flask of BGM cells. Two flasks were harvested at each time point, one to evaluate the yield of infectious virus and viral positive strand RNA, and the other to evaluate appearance of RF. Samples were taken at selected time points from 0 hr PI (end of adsorption period) until 44 hr PI. During the 44 hr incubation period, cells were observed for appearance of CPE. Before 28 hrs PI most cells appeared normal. At 32 hr PI, one area on the monolayer had 100% of cells rounded, which had the appearance of a large plaque on macroscopic observation of the flask. All rounded cells were attached. At 36 hr PI about 10-20% of the cells of the monolayer were rounded and the flask contained 2 of the larger foci or plaques with 50-100% of the cells rounded in the plaque area. At 44 hr PI, 99% of the cells were rounded, although all cells remained attached.

Comparison of yield over time of infectious virus in cells and supernatant and RT-PCR units in the cellular fraction is shown in Figure 3.2. The first cell-associated infectious CVB3 was detected at 8 hr PI, when no virus was detected in the supernatant fraction tested (0.5 mL of 5.0 mL total). The yield of infectious virus increased over time. A large increase in infectious virus production occurred between 8 and 32 hr PI. Total PFU per flask, RT-PCR units per flask, and PFU contained in the supernatant fluid all increased over time (Figure 3.2).

Total RT-PCR units per flask were estimated from an end-point titration of the same preparation that was measured in the plaque assay. Serial ten-fold dilutions were made of the chloroform-extracted cell-associated viral preparations, and RNA extracted from the dilutions with the QIAamp viral column method. The end-point titration, the volume assayed, and total volume of the flask preparation were used to calculate estimated total RT-PCR units per flask. Results are shown in Figure 3.2. There is an apparent increase over time in production of total infectious virus as well as total RT-PCR units per flask. At 36 hr PI and 44 hr PI the yield of RT-PCR units and total cell PFU is similar.

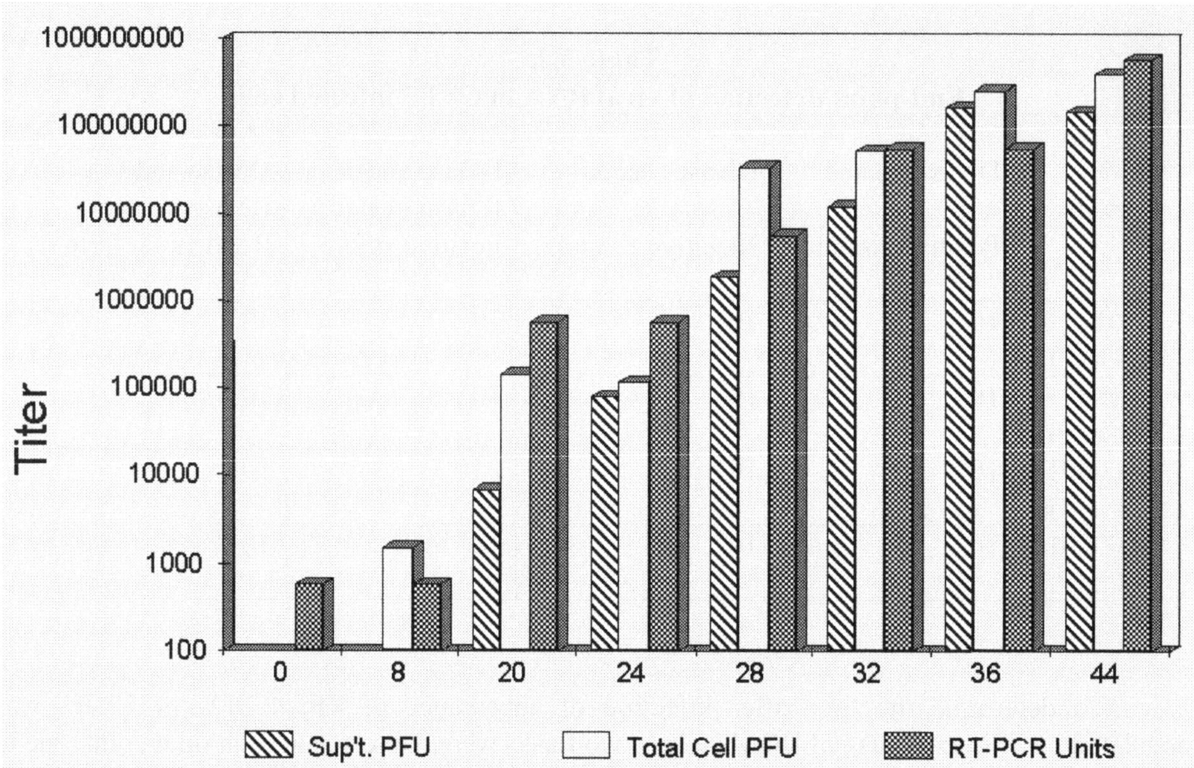

Figure 3.2. Comparison of infectious CVB3 in infected BGM cells and supernatant with RT-PCR units in infected cells over time after a low MOI infection

At the same times post infection, a second flask of cells was harvested with lysis buffer for RNA extraction and preparation of RF (containing negative and positive strands) with the enzyme treatments. Results of these analyses are shown in Figure 3.3. In order to compare appearance of RF with yield of infectious virus, the expected PFU was calculated at each time point for the volume assayed in the RT-PCR. Expected total PFU per flask was calculated from the duplicate flask. The PFU represented in the volume analyzed by RT-PCR were calculated from the fraction of the total volume of RNA per flask and the fraction of RNA actually analyzed per RF RT-PCR reaction. The end-point detections after assay for positive sense and negative sense RNA are roughly equal, indicating that the double stranded replicative form is detected. At 20 hr PI, infectious virus and RF positive and negative strands are detected. Definitive detection of RF was achieved at 20-24 hr PI in this experiment. RF were detected before any significant CPE was detected. Also at 24 hr PI, RNase treated RNA was assayed with no primers in the RT reaction to check for background at dilutions of 10^1, 10^2 and 10^3, and no PCR products were obtained. This indicates that no non-specific priming occurred with any remaining nucleic acid (small DNA fragments) after the nuclease treatments. In contrast, when specific primers were used in the RT reaction, specific product was obtained with dilution as high as 10^4. This indicates that the product obtained is specific for RF detection. In summary, as early as 24 hr PI, even at the very low infectious input of 4 PFU per T25 cm^2 , RF are detectable, are at peak production by 28 hr PI and remain detectable until 44 hr PI when CPE is finally present in 99% of cells.

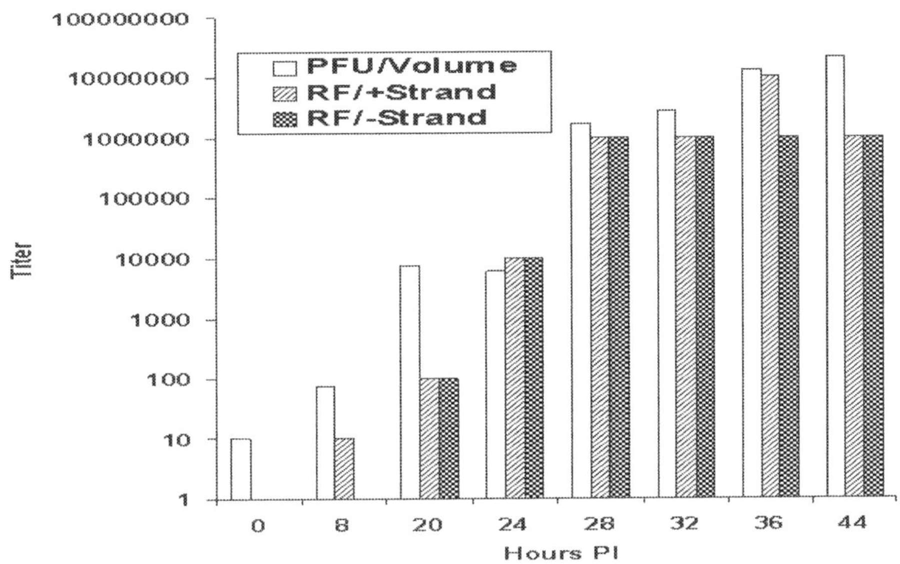

Figure 3.3. Kinetics of RF and infectious CVB3

Because disinfected viruses from treated water may be found in some source waters, it is important to prove the potential effectiveness of the RF detection method for discrimination of infectious and inactivated/non-infectious virus. A chlorine-disinfected inoculum (described in materials and methods) was evaluated in the assay. No infectious virus or RF were produced over the 35 hr incubation period (Table 3.3). In contrast, when the same quantity of untreated CVB3 was inoculated and evaluated for total RT-PCR units and RF production at the time points, the expected virus genomes and RF were produced. The finding of 290 viral positive strand RT-PCR units in the cells inoculated with disinfected virus, which is 10-fold more than that found in the cells inoculated with untreated virus at the end of adsorption (0 time), may be interpreted in at least two ways. One interpretation is that virus could be adsorbed to the cells and remains detectable in the RT-PCR. Another interpretation is that there is a nonspecific product formed in the reaction produced from the total RNA from the cells, which is primed with the positive sense primer and not the negative. Despite the RT-PCR amplification of RNA from the chlorinated virions, the results of this experiment demonstrate that RF detection is specific for infectious virus, because no amplicons were detected by RT-PCR of RF/negative strand.

Table 3.3
Evaluation of disinfected vs. non-disinfected CVB3 for detection of total virus RNA production and RF

Time PI	Disinfected CVB3 inoculum-- RT-PCR units per flask	RF/positive strand End-point	RF/negative strand End-point	CVB3 untreated inoculum- RT-PCR units per flask	RF/ Positive strand End-point	RF/Negative strand End-point
0 hr	2.9×10^2	10^1	Neg.	2.9×10^1	10^1	10^1
8 hr	NT	NT	NT	2.8×10^5	10^1	10^1
20 hr	2.9×10^2	10^1	Und	2.8×10^6	10^3	10^3
24 hr	2.9×10^2	10^1	Und	2.8×10^6	10^4	10^3
28 hr	NT	NT	NT	2.8×10^8	10^4	10^4
35 hr	2.9×10^2	Und.	Und	2.8×10^9	10^5	10^6

Replication of wild-type HAV in selected cells lines and evaluation of RF kinetics

Analysis of the potential for wild-type HAV replication and the ability to detect its infectivity by RF production was investigated in this study. Wild-type strains of HAV generally grow poorly if at all, and without CPE in cell culture (Hollinger and Emerson, 2001). The infectivity of HAV in cell cultures is low compared to human or primate infectivity and relative to the number of virions. Different cell lines were evaluated with two different HAV strains in an attempt to determine the best conditions for detection of replication and production of RF by wild-type HAV. HAV wild-type virus preparations from stool extracts, strains Q#1 and HLD-2, were inoculated onto FRhK-4 cells. After 4 weeks in culture with weekly feeding of the cells, one half of the monolayer was scraped into DPBS and prepared as virus stock, from which RNA was extracted for viral genome RNA detection by RT-PCR. The other half of the monolayer was lysed with Ambion Totally RNA lysis buffer and stored at –70°C for preparation and analysis of RF RNA. Results of the analysis for viral RNA only (no RF analysis was performed in this particular experiment) are shown in Table 3.4. After one month, the quantity of virus in the cells, based on viral RNA detection, was low. Virus positive strand RNA detected in the supernatant at four weeks PI could be due to some virus replication and perhaps slow release from cells or cell degeneration. Any non-encapsidated (mRNA) present would be detected also, but is unlikely present since the cells were not lysed. The cultures were refed weekly and this supernatant is only a one-week accumulation from week 3 to week 4. Other supernatants were not collected at the weekly refeeding during these experiments. Another explanation for the detection of HAV RNA in the supernatant is that virus was adsorbed, not uncoated and slowly released from the cells. The early steps of HAV replication including adsorption and uncoating of RNA are not well understood, and have not been well described (Hollinger and Emerson, 2001).

Table 3.4
Analysis of viral RNA one month after infection with wild-type HAV

HAV inoculum	Total input RT-PCR units	RT-PCR units total/cells	RT-PCR units total/supernatant
HLD-2	2.0×10^7	5.7×10^5	$\geq 7.0 \times 10^3$
Q#1 undiluted-A	4.3×10^4	5.7×10^2	5.0×10^1
Q#1 undiluted-B	4.3×10^4	5.7×10^2	5.0×10^1
Q#1 1:10	4.3×10^3	4.0×10^1	5.0×10^1

A cellular receptor for HAV has been isolated from cultured monkey kidney cells and a smaller receptor glycoprotein has been isolated from human liver and kidney cell lines (Fiegelstock et al. 1998). The human homologue receptor is expressed in many organs with higher levels in the cells of kidneys and testes than in liver. If the cellular receptor is present on many cell types including those in which the virus does not replicate, it is possible that HAV does attach to cells and remain without uncoating and replication.

In order to evaluate the possibility that measurement of virus in the cells represents input virus adsorbed and/or uncoated after one month of incubation time, a virus adsorption study was done. Results are shown in Table 3.5. These results suggest that very little virus was adsorbed to the cells, because 98-99.99% of virus was detected in the inoculum that was removed and/or in the cell wash after adsorption. The quantity of virus adsorbed to the cells is only slightly less than that present after one month in culture. This data suggests that the increase in titer seen at passage level 1 represents replication of virus, although further proof is required.

Table 3.5
Adsorption of wild-type HAV to FRhK-4 Cells as determined by viral RNA RT-PCR units

Wild-type HAV	Inoculum Total RT-PCR units	Inoculum removed-Total RT-PCR units	Wash-Total RT-PCR units	Cells Total RT-PCR units
Q#1	4.3×10^4	2.1×10^4	7.1×10^3	7.1×10^2
HLD-2	2.0×10^7	1.4×10^5	$\geq 2.8 \times 10^7$	7.1×10^4

The stability of wild-type HAV preparations at 37 °C for one month was determined. The Q#1 stock undiluted and diluted 1:10 in DPBS was stored and end-point RT-PCR titers determined at the end of the month. The RT-PCR titer did not decrease (results not shown). This suggests that measurement of virus in the cells without an increase in titer could be measurement of input virus, even after one month, if virus was adsorbed to cells and RNA was not degraded. Evidence of virus replication cannot be unambiguously established based on results obtained by direct RT-PCR of RNA from inoculated cell cultures. Analysis for production of RF was undertaken to determine if HAV replication was occurring. This data is presented later (Table 3.10).

The virus preparations that were analyzed for HAV genome RNA were also used to inoculate cells for the next virus passage. Serial passage of the HLD-2 and Q#1 is shown in Table 3.6.

Table 3.6

Analysis of viral RNA at passage levels 1-3 of wild-type HAV

HAV inoculum	Inoc. RT-PCR units for p.1	RT-PCR units total/cells p.1	RT-PCR units inoculated for p. 2	RT-PCR units Total /cells p.2	RT-PCR units inoc. for p.3	RT-PCR units Total/ cells p.3
HLD-2	2.0 $x10^7$	5.7 $x10^5$	flsk A. $7x10^4$	5.7 $x10^6$	$7x10^5$	$5.7x10^7$
Q#1	4.3 $x10^4$	5.7 x 10^2	flsk A. $2x10^2$	5.7 x 10^2	$7x10^1$	<10
		5.7 x 10^2	flsk B. $2x10^2$	5.7 x 10^2	$7x10^1$	$7x10^1$

Because only 1% or less of viral inoculum, based on RT-PCR units, was adsorbed during the first passage level, and the unadsorbed inoculum was removed after the 90-minute adsorption period (Table 3.5), the increase in HAV RT-PCR units for strain HLD-2 during the four weeks of each passage level is probably greater than the apparent 80-fold increase compared to the inoculum RT-PCR titer. At each passage level, the HLD-2 titer increased approximately ten-fold. Likewise, based on RT-PCR titers of the initial inoculum and an adsorption efficiency of much less than 1%, the first passage of Q#1 also could be interpreted as replication. However, this was not sustained at passage levels 2 and 3. More important in the interpretation of results, is that analysis of RF was negative in the Q#1 inoculated samples.

Further evaluation of HLD-2 replication in different cell lines based on detection of RF is presented in Table 3.7. Results of three separate trials in FRhK-4 cells are presented for comparison with initial trials in the other cell lines. Additional data on the HLD-2 replication in FRhK-4 cells is presented with seven serial passages described in a latter section of the document. The total RT-PCR units recovered from a flask of infected Caco2 cells at passage level 1 was similar to that recovered from a flask of FRhK-4 cells infected at passage level 1, which was the beginning of successful propagation of the virus in FRhK-4 cells. However, the second passage in Caco2 cells yielded a decreased titer. Similarly the RT-PCR titer at first passage in MA-104 cells was similar to that seen in FRhK-4 cells, however the RT-PCR titer at passage level 2 decreased significantly. From these data on detection of RT-PCR viral units, replication of wild-type HAV could have occurred initially in the first passage. However, for the cell lines other than FRhK-4, it was not sustained and there was no RF detection in these same cultures.

Table 3.7
Summary of wild-type HAV, HLD-2 RNA titers of different RNA species in three different cell lines

Cell Line Passage level	Virus passage Number and Trial	Total RT-PCR units per flask [A]	RF Negative strand end-points	RF Positive strand end-points
FRhK-4, ~ 90	1A	5.7×10^5	10^1	10^2
85	1B	5.7×10^4	10^1	10^2
81	1C	2.8×10^5	10^2	10^2
Caco2, 34	1A	5.7×10^5	Negative	Negative
34	1B and C	5.7×10^4	NT[B]	NT
24	2A, B and C	$<7 \times 10^2$	NT	NT
43	2A, B and C	5.7 X 10	NT	NT
PLC-PRF 5, 92	1	5.7×10^2	Negative	Negative
99	2	5.7×10^1	NT	NT
MA-104, 155	1A	5.7×10^5	Negative	Negative
155	1B	5.7×10^4	Negative	Negative
171	2A	5.7×10^2	Negative	Negative
171	2B	1.4×10^3	Negative	Negative
82	2A	1.4×10^3	NT	NT
82	2B	5.7×10^2	NT	NT
182	3A and B	<57	NT	NT

[A] Calculated from end-point titration of a portion of the total volume of the flask of cells.
[B] Not tested

The Q#1 strain of HAV was also inoculated into the various cell lines and followed for production of different RNA species as evidence of virus replication over several passages. Results of these experiments are shown in Table 3.8. Passage level 1 and 2 of Q#1 in FRhK-4 cells could be interpreted as potential replication with the viral RNA analysis only, however, if this was virus replication, it was not maintained when passaged through level 5, and no RF were detected in passage 2. The infectivity of Q#1 inoculum is not known, whereas the infectivity of HLD-2 has been established in an animal model. The RT-PCR units per volume of the Q#1 are at least 1000-fold lower than the HLD-2 material. Two attempts to initiate an HLD-2 infection with a 1000-fold dilution of the stock virus were unsuccessful (data not shown), suggesting that this lower level inoculum was inadequate to initiate infection under the conditions used and with a permissive cell line. The Q#1 preparation was not successful in initiating infection in any cell lines that were examined under the conditions used.

Table 3.8
Evaluation of wild-type HAV, Q#1 RNA production in different cell lines

Cell line Passage level		Virus passage (Inoc. Dilution)	Total RT-PCR virus units per flask	RF (negative strand assay and positive strand assay)
FRhK-4	~90	1	2.8×10^2	NT
	~90	2	2.8×10^2	Neg
	~90	3, 4 and 5	Neg	NT
Caco2	33	1A and B	2.8×10^3	Neg
	37	2A and B	Neg	NT
PLC PRF5	96	1A and B	2.8×10^2	Neg
	96	1 C (1:3 inoc)	2.8×10^1	NT
	96	2 A and B	Neg	NT
	92	3 A	Neg	NT
MA-104	155	1	5.7×10^3	NT
	171	2	5.7×10^1	NT
	82	2	Neg	NT
	182	3	Neg	NT

Successive passage of the HAV HLD-2 inoculum in FRhK-4 cells was confirmed by IC-RT-PCR examination of the preparations used for successive passage in addition to the RT-PCR examination. In this method, virus is captured on antibody coated microfuge tubes and after washes; RT-PCR is performed in the same tube. Only viruses with a capsid capable of immunological reaction with the capture antibodies on the walls of the tubes should be captured, retained and assayed with this method. Data in Table 3.9 demonstrate that IC-RT-PCR titers are 10 fold lower (passages 1, 2, and 3) or equal to the titer by RT-PCR (passage 4 and 5). These results provide additional supportive evidence that measurement of sustainable or increasing RT-PCR titers over serial passages is a measure of viral production.

Table 3.9
Comparison of RT-PCR and IC-RT-PCR quantitation of viral RNA in HLD-2 infected cells

Passage level	RT-PCR yield	IC-RT-PCR yield
1	2.8×10^5	2×10^4
2A	2.8×10^6	2×10^5
3A	2.8×10^6	2×10^5
4A	2.8×10^6	2×10^6
5B	6.0×10^6	2×10^6

The complete analysis of RNA species in HLD-2 HAV inoculated FRhK-4 cells during serial passages is shown in Table 3.10. Seven successive passages of HLD-2 demonstrated the successful replication of virus based on detection of various HAV RNA species, including RF. To confirm the presence of RF a second round of PCR was performed. Results in the two right hand columns demonstrate that a second round of PCR analysis of these samples yielded confirmatory results for most end-points. Weekly analysis of the RF and total virus RNA production was performed to determine kinetics of production and potential loss of RF during the 4 weeks of culture. The first inoculation and passage was performed three times (1.1 is the first passage the first time, 1.2 is the first passage the second time, and 1.3 is the first passage performed the third time). Passage 1.1 (also shown as passage 1 in table 3.6) was not analyzed at weekly intervals, however the other two (1.2 and 1.3) were. All three first passages have RF present at four weeks PI, as confirmed by a second round (nested) PCR shown in the two left columns. However, detection of RF in early passage levels was not as consistent as at later passage levels. RF production was consistent and a higher titer at passage level 3A, when even at 3 weeks PI (see bold in Table 3.10) the highest level was seen. At least a ten fold increase over total RT-PCR titer (column 2) in the virus preparation was also obtained at passage level 3A. In general RT-PCR units per flask increased with each week of incubation over the four week period, however, in some instances there is a decrease in titer, such as week two to three in 1.3 passage and week three and four in passage 4A. This could be explained by variation in the assay or by actual differences due to the different flasks of cells. We did observe cells for consistency throughout the experiments. The highest total RT-PCR titers were obtained at passage seven when the passage 6 virus diluted 1:10 was the inoculum. At week 4 of passage 7A, the titer is equal to that obtained with all well-established cell culture-adapted strains of HAV, such as HM-175.

Table 3.10
Evaluation of wild-type HLD-2 replication for total genome RNA production and RF production

Passage level and week PI	Total Yield of RT-PCR units per flask	RF/positive strand reaction End-point	RF/negative strand reaction End-point	Nested of positive strand RF reaction	Nested of negative strand RF reaction
1.1 Wk4	5.7×10^5	10^2	Und	10^2	Und
1.2 Wk 1	5.7×10^5	Und	Und	10^1	10^1
Wk 2	5.7×10^5	Und	Und	10^1	10^1
Wk 3	5.7×10^5	10^2	10^2	Und	Und.
Wk 4	5.7×10^4	10^2	10^1	10^2	10^1
1.3 Wk 1	2.8×10^5	NT	NT	10^1	10^2
Wk 2	1.4×10^4	NT	NT	Und.	Und
Wk 3	2.8×10^5	NT	NT	10^1	10^1
Wk 4	2.8×10^5	NT	NT	10^2	10^2
2A Wk 1	2.9×10^2	Und	Und	Und	Und
Wk 2	2.9×10^3	10^1	Und	Und	Und
Wk 3	1.4×10^4	10^1	10^1	10^1	10^1
Wk 4	2.9×10^4	Und	10^{-1}	Und	10^1
3A Wk 1	2.9×10^5	Und	Und	10^1	Und
Wk 2	2.9×10^5	10^1	10^2	10^1	10^2
Wk 3	$\mathbf{1.4 \times 10^6}$	$\mathbf{10^2}$	$\mathbf{10^2}$	$\mathbf{10^2}$	$\mathbf{10^2}$
Wk 4	2.9×10^6	10^3	10^2	10^3	10^2
4A Wk 1	2.9×10^5	Und.	10^{-1}	10^1	10^2
Wk 2	1.4×10^6	Und.	10^1	10^1	10^1
Wk 3	2.9×10^7	Und.	10^1	10^3	10^1
Wk 4	2.9×10^6	10^1	10^2	10^2	10^3
5A Wk 1	6.0×10^4	Und	Und	Und	Und
Wk 2	6.0×10^5	10^3	10^{-4}	10^3	10^4
Wk 3	6.0×10^6	10^2	10^{-4}	10^2	10^4
Wk 4	6.0×10^7	10^2	10^{-3}	10^2	10^3
6A Wk 4	1.4×10^7	10^3	10^{-3}	10^3	10^3
6B Wk 4	2.9×10^6	10^3	10^{-2}	10^3	10^2
7A (1:10) 1	1.4×10^5	Neg	10^2	10^1	10^2
Wk 2	2.9×10^7	10^3	10^3	10^3	10^3
Wk 3	2.9×10^7	10^3	10^3	10^2	10^3
Wk 4	$\mathbf{1.4 \times 10^8}$	$\mathbf{10^3}$	$\mathbf{10^4}$	$\mathbf{10^3}$	$\mathbf{10^4}$

All the data indicate that in HLD-2 HAV infection, RF production occurs and can be detected by the methods developed. However, RF is more difficult to detect in early passages, and occurs at higher levels as virus is serially passaged and adapted to cell culture conditions. The non-detectable or low RF titers of HAV in early passages and the increase with increasing passage is consistent with the widely observed phenomenon that many wild-type viruses adapt to cell cultures during serial passage. RNA replication has been suggested as one of the several potential reasons for the slow replication of HAV in cell culture (Hollinger and Emerson, 2001). The results observed for RF production in these experiments are consistent with that hypothesis.

The data we have obtained from these experiments indicate that culture of wild-type HAV is very difficult, and when achieved, it is relatively inefficient. These results are consistent with previous studies in which there was only a low level presence of negative strand in HAV infected cells, based on detection by immunofluorescence methods, in comparison to poliovirus infected cells (Anderson, Ross, and Locarnini 1988). The only clearly successful replication of a wild-type HAV strain was obtained with a high titered inoculum of a strain of HAV (HLD-2) known to be infectious in primates, and it was serially propagated in only one of the four cell lines tested. Nevertheless, the methods developed and used in this study demonstrate that RF (double stranded RNA) is present early in infection with the wild-type virus and this provides a definitive basis for detecting the infectivity of some wild-type HAV strains in some cell cultures at least some of the time.

RF production and detection as evidence of wild-type enterovirus infectivity

Wild-type enteroviruses were isolated from a sewage sample as described in methods and materials. Eleven isolates were obtained by plaque isolation. Each wild-type virus plaque isolate (extracted by chloroform from the agarose sample) was assayed to determine the infectivity titer. The days required for plaque formation in the assay were longer than required for cell culture-adapted CVB3 virus (3-5 days vs. 2 days). The titers ranged from 1×10^4 to 3.8×10^6 PFU/ml. Two of these virus isolates were identified by sequence analysis of virus from a second passage in BGM cells in the enterovirus laboratory of Dr. Steve Oberste at CDC (Oberste et al., 1999A, 1999B, and 2000). With these methods they were identified as Echovirus 1 and CVB4.

Replication kinetics of RNA species of these two identified wild-type viruses were determined using the second passage material as the cell culture inoculum. BGM cells were infected with 4 PFU/T25 cm^2 flask by the same method as previously described for the CVB3 experiments. Separate flasks were analyzed for total RT-PCR units and for RF production. Results from infection with Echovirus 1 are shown in Table 3.11 and from infection with CVB4 are shown in Table 3.12.

Table 3.11
Detection of wild-type Echovirus 1 and RF in BGM cells

Hrs Post Infection	Total RT-PCR units	RF/positive strand end-point	RF/negative strand end-point
0	$<1.4 \times 10^1$	ND	ND[a]
12	2.8×10^2	10^1	10^1
24	2.8×10^4	10^2	10^3
36	2.8×10^6	10^2	10^3
48	1.4×10^8	10^3	10^3
54	1.4×10^9	10^3	10^3
60	1.4×10^8	10^3	10^3

[a]ND, not detected

Small foci of infection could be seen macroscopically and confirmed microscopically by 36 hr PI in Echo1 infected cells and somewhat later in CVB4 infected cells. By 60 hr PI, 100% of Echo 1 infected cells were rounded and some were even detached and possibly lysed. In contrast the CVB4 infected cells were 100% attached although rounded with CPE at 72 hr PI. Although the rate of replication was related to the initial titers and time required for isolation by plaque detection, both yielded similar titers at 48 to 72 hr PI. As shown in tables 3.11 and 3.12, RF production was clearly evident by 24 hr PI in Echo 1 infected cells and 36 hr PI in CVB4 infected cells. These data indicate that RF production in wild-type virus infected BGM cells can be minimally detected one day post infection and readily detected two to three days after inoculation with very low levels of virus, even if replication is not as efficient as that of laboratory-adapted strains of viruses in these cells.

Table 3.12
Detection of wild-type CVB4 and RF in BGM cells

Hrs Post Infection	Total RT-PCR units	RF/Positive strand end-point	RF/Negative strand end-point
0	1.9×10^1	Und.	Neg.
24	2.8×10^4	10^1	10^1
36	2.8×10^5	10^3	10^3
48	2.9×10^6	10^4	10^4
54	1.4×10^8	10^4	10^4
60	5.8×10^8	10^4	10^5
72	2.9×10^8	10^5	10^5

Practical application of the method developed in this project for detection of RF of the virus in inoculated cultures was confirmed by the study of two viruses isolated directly from concentrated sewage. Viruses isolated on BGM cells were evaluated at only the second laboratory passage for formation of detectable RF in cell culture. RF were detectable at 24 and 36 hr after initial infection of cultures with only 4 infectious units of Echovirus 1 and CVB4, respectively. These time points are about 24 hr before maximum infectious virus production, and before CPE could be seen. These experiments demonstrate the potential usefulness of the RF RT-PCR in rapid detection of wild-type enteroviruses that replicate in BGM cells.

DETECTION OF INFECTIOUS ADENOVIRUS BY mRNA ASSAY IN INOCULATED CELL CULTURES USING RT-PCR

Development and optimization of PCR and mRNA RT-PCR

New primers and previously described primers were selected for use in this study to perform PCR and mRNA RT-PCR of adenoviruses in cell cultures as evidence of virus infectivity. Initially, the PCR primers used in this study were evaluated to identify optimal assay conditions. Detection sensitivities are summarized in Table 3.13. When PCR was performed under optimized conditions with heat-release of Ad genomes, as few as 1 IU of Ad2 was detected with the AdC-E1AF/AdC-E1AR primer pair. As few as 10 IU of Ad2 were detected by PCR with Hex1/Hex2 primers and as little as 0.1 IU of Ad2 was detected by subsequent hemi-nested PCR with Hex1/Hex3 primers. PCR with AdF-E1AF/AdF-E1AR primers detected as low as 0.01 IU of Ad41. PCR with Hex1/Hex2 primers and subsequent hemi-nested PCR with Hex1/Hex3 primers detected as low as 1 IU and 0.01 IU of Ad41, respectively.

Table 3.13
Sensitivity of adenovirus detection by standard and hemi-nested PCR using candidate primers

Ad types	E1A primers[a]	Hexon primers[b]	
	PCR	PCR	Hemi-nested PCR
Ad 2	1 IU	10 IU	0.1 IU
Ad 41	0.01 IU	1 IU	0.01 IU

[a] PCR with AdC-E1AF/AdC-E1AR primers for Ad2 and AdF-E1AF/AdF-E1AR primers for Ad41
[b] PCR with Hex1/Hex2 primers and nested PCR with Hex1/Hex3 primers.

Kinetics of Ad2 and Ad41 infection detected by RT-PCR of mRNA

When Ad2 was inoculated into A549 cells at a multiplicity of 0.1 IU per cell, E1A mRNA was detected as early as 6 hr PI and hexon mRNA was detected as early as 24 hours PI (Figure 3.4). RT-PCR without reverse transcriptase (RTase) indicated that no viral DNA contamination remained in the mRNA extracts because no amplicon was detected. Therefore, the Ad2 mRNA RT-PCR using the E1A primer set gives sensitive, rapid and specific detection of infecting virus. Ad2 mRNA RT-PCR using the hexon primer set also was sensitive and specific, although slower to produce results.

(A) E1A primer set

	M	1	2	3	4	5	6	7	8	9	10	11	12	13	14
Result:		-	+	+	+	+	-	-	-	-	-	-	-	-	+

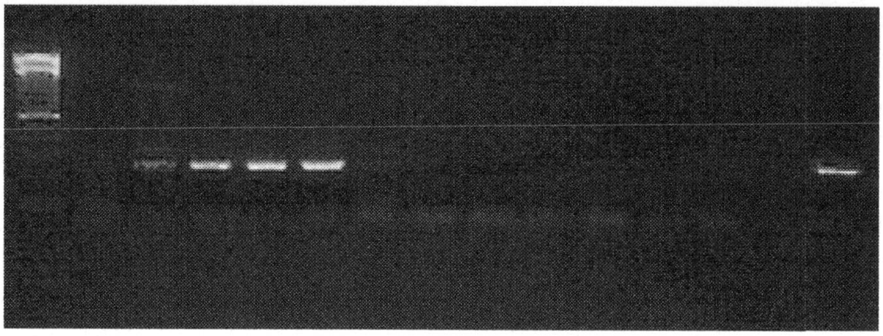

←260bp

(B) hexon primer set

	M	1	2	3	4	5	6	7	8	9	10	11	12	13	14	15	16	M
Result:		-	-	+	+	+	-	-	+	-	-	-	-	-	-	-	+	

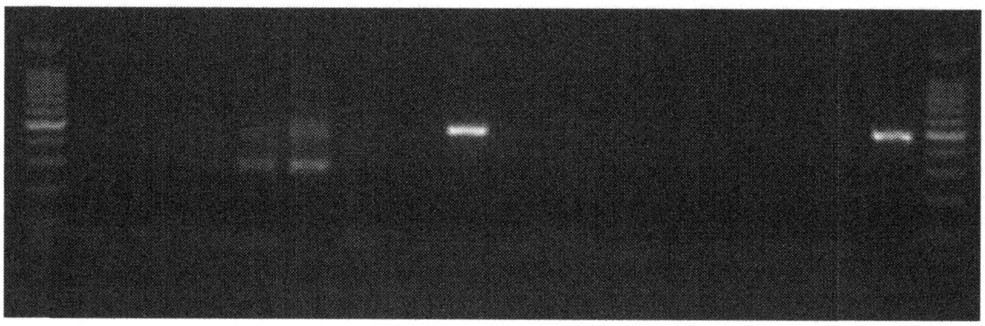

←482bp

Figure 3.4. Detection of Ad2 mRNA by RT-PCR at different times PI. mRNA was extracted from A549 cells infected with Ad2 at 0.1 IU/cell. (A) lanes 1 to 5, RT-PCR of mRNA extracted at 0, 6, 18, 24, 36 hour PI; lane 6, RT-PCR of cell control; lanes 7-11 RT-PCR without RTase of mRNA extracted at 0, 6, 18, 24, 36 hour PI; lane 12, RT-PCR without RTase of cell control; lane 13, negative control; lane 14, positive control. (B) lanes 1 to 5, RT-PCR of mRNA extracted at 0, 6, 18, 24, 36 hour PI; lane 6, RT-PCR of cell control; lane 7, negative control; lane 8, positive control; lanes 9-14, RT-PCR without RTase of mRNA extracted at 0, 6, 18, 24, 36 hour PI; lane 15, negative control; lane 16, positive control. A 100-bp DNA ladder was used as a molecular weight marker (Lane M).

When Ad41 was inoculated into G293 cells at a multiplicity of 0.1 IU per cell, E1A mRNA was detected after 7 days PI, which is much later than the detection of Ad2 MRNA in A549 cells (Figure 3.5A). However, under the same conditions of infection, hexon mRNA was detected as early as 24 hours PI and remained detectable through 7 days PI (Figure 3.5B). RT-

PCR without RTase indicated that no contaminating viral DNA remained in the mRNA extract because no amplicons were detected. Therefore, the Ad41 mRNA RT-PCR using the hexon primer set gives sensitive, rapid and specific detection of infecting virus. Ad41 mRNA RT-PCR using hexon primers also was specific and sensitive although considerably slower to provide results.

(A) E1A primer set

M 1 2 3 4 5 6 7 8 9 10 11 12 13 14 15 16 17 18 M

Result: - - - - - + - - + - - - - - - - - +

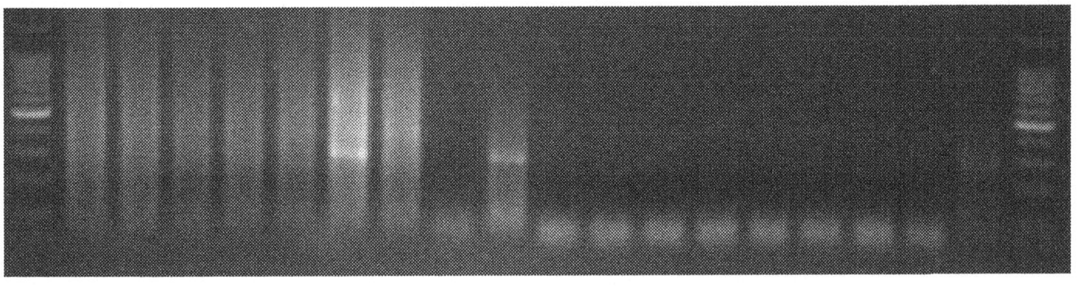

←280bp

(B) hexon primer set

M 1 2 3 4 5 6 7 8 9 10 11 12 13 14 15 16

Result: - - - + + + - - + - - - - - - -

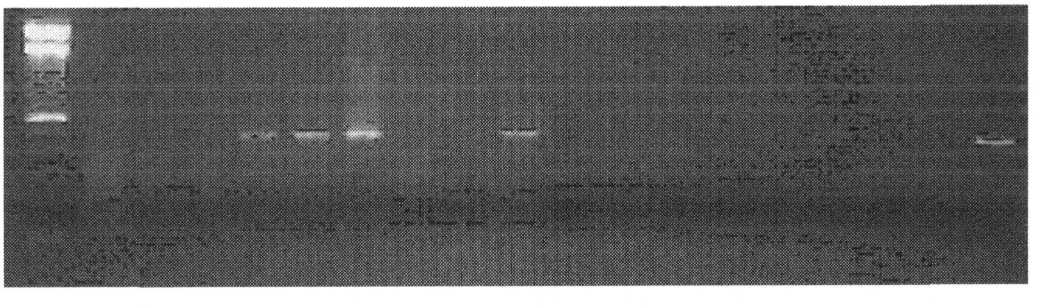

←482bp

Figure 3.5. Detection of Ad41 mRNA by RT-PCR at different times PI. mRNA was extracted from A549 cells infected with Ad2 at 0.1 IU/cell.. mRNA was extracted from G293 cells infected with Ad41 at 0.1 IU/cell. (A) lanes 1 to 6, RT-PCR of mRNA extracted at 0, 6, 12, 24, 36 hours and 7 days PI; lane 7, RT-PCR of cell control; lane 8, negative control; lane 9, positive control, lanes 10-15 RT-PCR without RTase of mRNA extracted at 0, 6, 12, 24, 36 hours and 7 days PI; lane 16, RT-PCR without RTase of cell control; lane 17, negative control; lane 18, positive control. (B) lanes 1 to 6, RT-PCR of mRNA extracted at 0, 6, 12, 24, 36 hours and 7 day PI; lane 7, RT-PCR of cell control; lane 8, negative control; lane 9, positive control, lanes 10-15 RT-PCR without RTase of mRNA extracted at 0, 6, 12, 24, 36 hours and 7 days PI; lane 16, RT-PCR without RTase of cell control. A 100-bp DNA ladder was used as a molecular weight marker (Lane M)

Detection of Ad41 mRNA by RT-PCR using standard and hemi-nested methods at different times PI in cell cultures inoculated with different amounts of virus is shown in Figure 3.6. Approximately 10^7 cells of G293 were infected with different quantities (5×10^4, 5×10^2, 50, 5 IU) of Ad41. Infected cells were harvested at 1, 2, 3, 5 days PI, and then mRNA extracts from infected cells were tested by RT-PCR with and without RTase using Hex1 and Hex2 primers. When cells were infected with 5×10^4, 500, and 50 IU, hexon mRNA was detected as early as 1, 3, and 5 days PI by mRNA RT-PCR, respectively. Hexon mRNA was not detectable by the first round of RT-PCR when cells were infected with only 5 IU. However, greater detection sensitivities were observed when hemi-nested PCR was performed. In cells infected with 5×10^4, 5×10^2 IU of Ad41, hexon mRNA was detected within 1 day by hemi-nested RT-PCR. Hexon mRNA also was detected by hemi-nested RT-PCR as early as 3 days post-inoculation in cells infected with 5 or 50 IU of Ad41. The results of these experiments show that low levels of Ad41, as low as 5 IU, can be detected by hemi-nested RT-PCR in as little as three days. Standard RT-PCR was less sensitive than hemi-nested RT-PCR, with no viral mRNA detected when the inoculum was 5 IU but positive detection at five days when the inoculum was 50 IU.

(A) RT-PCR

M 1 2 3 4 5 6 7 8 9 10 11 12 13 14 15 16 17 18 M

Result: - + + + - - + + - - - + - - - - - +

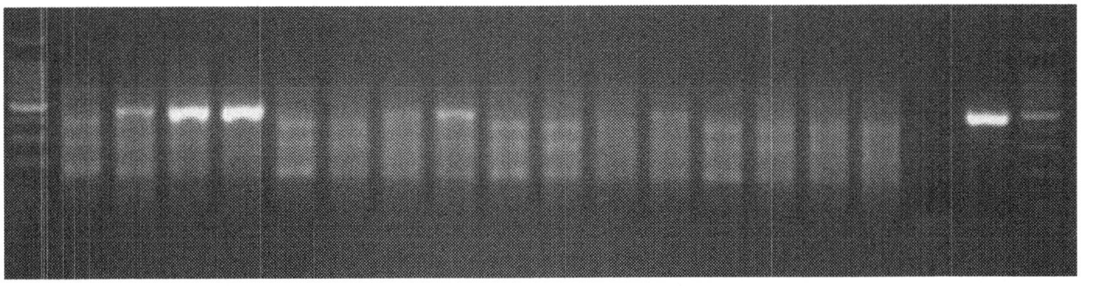

←482bp

(B) Hemi-nested PCR

M 1 2 3 4 5 6 7 8 9 10 11 12 13 14 15 16 17 18 M

Result: - + + + - + + + - - + + - - + + - +

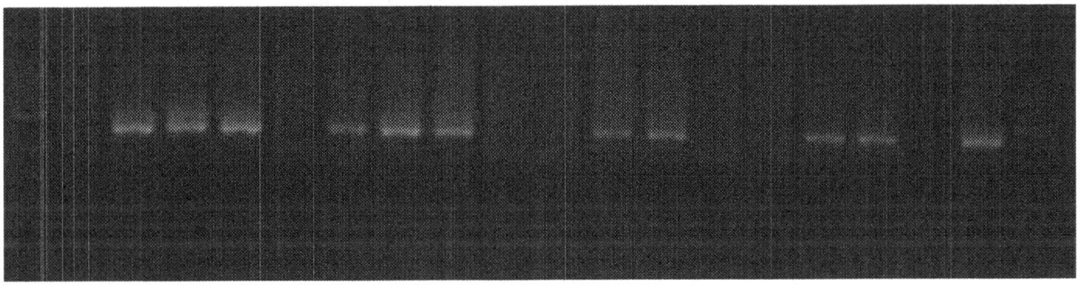

←443bp

Figure 3.6. Detection of Ad41 mRNA by RT-PCR and subsequent nested PCR by hexon primer sets at different times PI. (A) lanes 1 to 4, RT-PCR of mRNA from G293 cells infected with 5×10^4 IU of Ad 41 at 0, 1, 3, 5 days PI; lanes 5-8, RT-PCR of mRNA from G293 cells infected with 5×10^2 IU of Ad 41 at 0, 1, 3, 5 days PI; lanes 9-12, RT-PCR of mRNA from G293 cells infected with 50 IU at 0, 1, 3, 5 days PI; lanes 13-16, RT-PCR of mRNA from G293 cells infected with 5 IU at 0, 1, 3, 5 days PI; lane 17, negative control; lane 18, positive control. (B) lanes 1 to 4, nested PCR of mRNA from G293 cells infected with 5×10^4 IU of Ad 41 at 0, 1, 3, 5 days PI; lanes 5-8, hemi-nested PCR of mRNA from G293 cells infected with 5×10^2 IU of Ad41 at 0, 1, 3, 5 days PI; lanes 9-12, hemi-nested PCR of mRNA from G293 cells infected with 50 IU at 0, 1, 3, 5 days PI; lanes 13-16, hemi-nested PCR of mRNA from G293 cells infected with 5 IU at 0, 1, 3, 5 days PI; lane 17, negative control; lane 18, positive control. A 100-bp DNA ladder was used as a molecular weight marker (lane M).

Lower limit of detection sensitivity of Ad2 and Ad41 infection by mRNA RT-PCR

Figure 3.7 shows the results as RT-PCR products of mRNA at different levels of viral inoculation into cell cultures. Detection of Ad2 E1A mRNA by RT-PCR was as low as 2 IU of virus inoculum after 7 days PI (Figure 3.7A). Six different samples containing from 10^3 to 10^{-3} IU were inoculated to cell cultures, and mRNA was extracted after 7 days PI. Ad41 hexon mRNA was detectable at inoculum levels as low as 10 infectious units after 7 days PI by RT-PCR. Hemi-nested PCR detected hexon mRNA at inoculum levels as low as 0.1 infectious unit of Ad41 by 7 days PI (Figure 3.7B). Therefore, detection of low levels of Ad2 and 41 mRNA was possible by 7 days, especially by hemi-nested RT-PCR. Standard RT-PCR was somewhat less sensitive but detected viral mRNA at inoculum levels as low as 2 to 10 IU.

(A) Ad2

M 1 2 3 4 5 6 7 8 9 10 11 12
Result + + - - - + - - - - - +

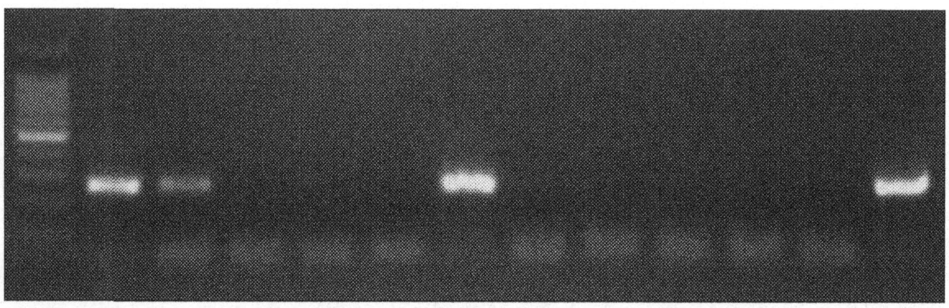

←260bp

(B) Ad41

M 1 2 3 4 5 6 7 8 9 10 11 12 13 14 15 16 17 18 M
Result: + + + - - - - - + - - - - - - - - +

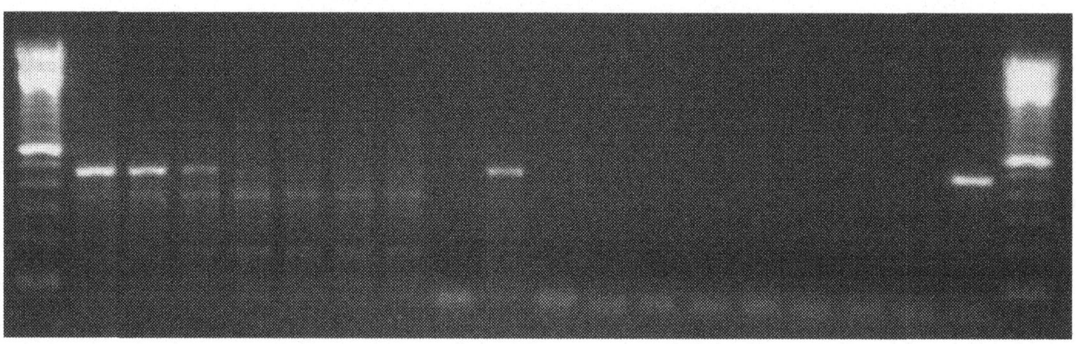

(continued)

Figure 3.7 (continued)

(B) Ad41
Hemi-nested PCR

M 1 2 3 4 5 6 7 8 9 10 11 12 13 14 15 16 17 18 M
Result: + + + + + - - - + - - - - - - - - +

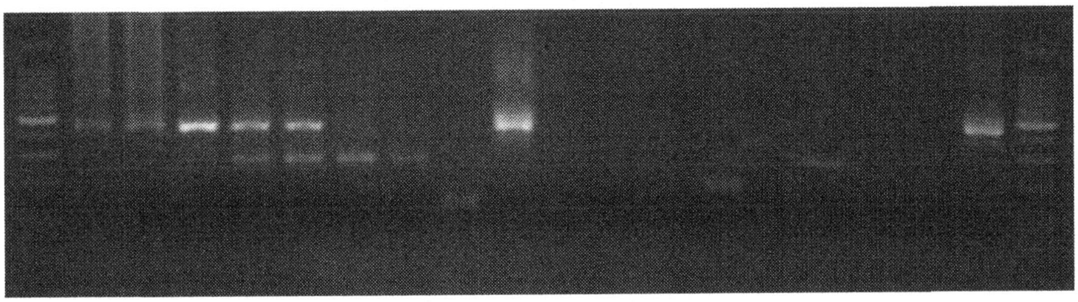

←443bp

Figure 3.7 (A and B). Sensitivity of detection of Ad2 and Ad41 mRNA at 7da PI. Ad2 mRNA was amplified with the E1A primer set and Ad41 mRNA with the hexon primer set. (A) lanes 1 to 3, RT-PCR of Ad2 mRNA from A549 cells infected with 20, 2, 0.2 IU; lane 4, RT-PCR of mRNA from non-infected A549 cells; lane 5, negative control; lane 6, positive control; lane 7 to 9, PCR of Ad2 mRNA from cells infected with 20, 2, 0.2 IU; lane 10, PCR of mRNA in non-infected A549 cells; lane 11, negative control; lane 12, positive control. (B) lanes 1 to 7, RT-PCR or subsequent hemi-nested PCR of Ad41 mRNA from G293 cells infected with 10^3, 10^2, 10, 1, 0.1, 0.01, 0.001 IU at 7 days PI; lane 8, negative control; lane 9, positive control; lanes 10 to 16, RT-PCR without RTase or subsequent nested PCR of mRNA from G293 cells infected with 10^3, 10^2, 10, 1, 0.1, 0.01, 0.001 IU at 7 days PI; lane 17, negative control; lane 18, positive control. A 100-bp DNA ladder was used as a molecular weight marker (lane M).

Ability of mRNA RT-PCR to distinguish between infectious and chlorine- or UV-inactivated Ad

Chlorine treated (free chlorine dose of 1 mg/l for 1, 10, 30 minutes in buffered water) Ad2 and Ad41 were inoculated into cell cultures as described. After 7 days PI, mRNA was extracted from inoculated cell cultures and assayed by mRNA RT-PCR. Non-chlorinated control viruses but not chlorinated viruses produced CPE in A549 cells (Table 3.14). No Ad2 mRNA was detected by E1A RT-PCR in cells inoculated with free chlorine-treated viruses, but Ad2 mRNA was detected in cells inoculated with untreated virus stock. Adenovirus DNA was detected by PCR in the virus inoculum even after the maximum dose of free chlorine treatment (Ct = 30 mg-min/L). These results indicate that direct PCR of Ad DNA targets can be detected in chlorine-inactivated viruses. Such PCR detection of chlorine-inactivated viruses can cause false-positive interpretations of adenovirus health risks in chlorine disinfected samples.

After 7 days PI, no CPE was observed in any G293 cells inoculated with Ad41. As found in Ad2 inoculated cells, no hexon mRNA was detected by RT-PCR in G293 cells inoculated with chlorine treated viruses, but hexon mRNA was detected by RT-PCR in cells inoculated with untreated virus stock. Increases in Adenovirus DNA concentrations were observed by PCR 7 days PI with untreated Ad2 and Ad41. These results for direct PCR of DNA indicate that increases in DNA over the time period of incubation must be detected in inoculated cell cultures in order to obtain some evidence that infectious viruses are present. The detection of viral DNA by PCR with no increase in DNA quantity or titer during the incubation period is not conclusive evidence of virus infectivity because the DNA of chlorine-inactivated viruses can be detected. Therefore, the detection of adenovirus DNA in cell cultures has to be interpreted with caution and must be done on a quantitative basis over time to obtain interpretable results regarding infectious virus.

Table 3.14.
Results of culture, PCR, and mRNA RT-PCR of Ad2 and Ad41 disinfected by different Ct doses of free chlorine.

(a) Ad2

	Free chlorine Dose as Ct (mg·min/L)[a]			
	0	1	30	100
CPE[b]	+	-	-	-
DNA detection (PCR)	+	+	+	+
Increased viral DNA during culture (PCR)	+	-	-	-
mRNA detection[c] (RT-PCR)	+	-	-	-

[a] Ct Dose = concentration (mg/L)×contact time (minute) in phosphate buffer (pH = 7.2) at 4 °C.

[b] Cytopathogenic effect on A549 cells after 10 days.

[c] mRNA was extracted from infected A549 cells at 10 days PI, and subjected to RT-PCR with AdC-E1AF/AdC-E1AR primers.

(b) Ad41

	Dose of free chlorine (Ct) (mg·min/L)[a]			
	0	1	30	100
CPE[b]	-	-	-	-
DNA detection (PCR)	+	+	+	-
Increased viral DNA during culture (PCR)	+	-	-	-
mRNA detection[c] (RT-PCR)	+	-	-	-

[a] Ct Dose = concentration (mg/L)×contact time (minute) in phosphate buffer (pH = 7.2) at 4 °C.

[b] Cytopathogenic effect on Graham 293 cells after 7 days.

[c] mRNA was extracted from infected G293 cells at 7 days PI, and subjected to RT-PCR with Ad1/Ad2 primers.

Ad2 (~5 x 10^5) and Ad 41 (~10^3) in PBS were exposed to different doses (0, 50, 100, 250 mJ/cm^2) of collimated, monochromatic (254 nm) UV radiation. Cells were infected as described in Methods and Materials. After 7 days PI, mRNA was extracted from inoculated cells and evaluated by mRNA RT-PCR and by PCR. As shown in Table 3.15, Ad2 viruses either untreated or exposed to lower doses (50, 100 mJ/cm^2) of UV produced CPE in A549 cells; Ad2 exposed to the highest UV doses of 250 mJ/cm^2 produced no CPE in these cells. After 7 days PI, no CPE was observed in any G293 cells inoculated with Ad41. Adenovirus DNA was detected by PCR in the virus inoculum even after the maximum dose of UV exposure. No mRNA was detected in cell cultures inoculated with Ad exposed to the highest UV dose (250 mJ/cm^2) whereas mRNA was detected in cell cultures inoculated with untreated virus (zero UV dose) and virus treated with the lower (50 and 100 mJ/cm^2) UV doses. For both Ad2 and Ad41 untreated and at the lower (50 and 100 mJ/cm^2) UV doses, increases in adenovirus DNA concentrations were observed by PCR after 7 days PI of cell cultures. The results of these UV disinfection studies indicate that the infectivity of adenoviruses can be detected conclusively and relatively rapidly by mRNA RT-PCR, and that there is no such detection when the virus is effectively inactivated using a high dose of UV radiation. The detection of infectious adenoviruses at the lower UV doses of 50 and 100 mJ/cm^2 by mRNA RT-PCR is consistent with the results of previous studies reporting that adenoviruses are not extensively inactivated by UV doses as high as 100 mJ/cm^2 (Gerba, Gramos, and Nwachuku, 2002).

Table 3.15
Results of culture, PCR, and mRNA RT-PCR of Ad2 and Ad41 disinfected by different doses of 254 nm ultraviolet radiation.

(A) Ad2

	Dose of UV (mJ/cm^2)[a]			
	0	50	100	250
CPE[b]	+	+	+	-
DNA detection (PCR)	+	+	+	+
Increased viral DNA during culture (PCR)	+	+	+	-
mRNA detection[c] (RT-PCR)	+	+	+	-

[a] Dose = UV irradiance (mW/cm^2)×exposure time (sec) in PBS (pH = 7.2) at room temperature.
[b] Cytopathogenic effect on A549 cells after 10 days.
[c] mRNA was extracted from infected cells at 10 days PI, and subjected to RT-PCR with AdC-E1AF/AdC-E1AR primers.

(continued)

Table 3.15 (continued)

(B) Ad41

	Dose of UV (mJ/cm^2)[a] 0	50	100	250
CPE[b]	-	-	-	-
DNA detection (PCR)	+	+	+	+
Increased viral DNA during culture (PCR)	+	+	+	-
mRNA detection[c] (RT-PCR)	+	+	+	-

[a] Dose = UV irradiance (mW/cm^2)×exposure time (sec) in PBS (pH = 7.2) at room temperature.

[b] Cytopathogenic effect on Graham293 cells after 7 days.

[c] mRNA was extracted from infected G293 cells at 7 days PI, and subjected to RT-PCR with Ad1/Ad2 primers.

Validation of mRNA RT-PCR assay for infectious Ad detection in environmental water sample concentrates.

The results of experiments to determine the ability of mRNA RT-PCR to detect infectious adenoviruses in environmental water sample concentrates are shown in Figure 3.8. Environmental water sample concentrates (obtained by PEG precipitation as described in Methods and Materials) seeded with relatively low numbers of Ad (IU per sample is shown in the Figure 3.8 legend) were inoculated onto cell cultures. After 1 hour of adsorption, the inoculum was removed, and cultures incubated for 7 days. Messenger RNA was extracted from cultured cells and analyzed by RT-PCR. As shown in Figure 3.8 (A), RT-PCR of Ad2 E1A mRNA detected as few as 2 IU in the NC water sample concentrate, and as few as 10 IU in the GA water sample concentrate. Hemi-nested RT-PCR of Ad41 hexon mRNA detected as few as 10 IU in the GA water sample concentrate (Figure 3.8 B). Therefore, low levels of infectious Ad2 and Ad41 can be detected by RT-PCR of mRNA from inoculated cell cultures at 7 days PI.

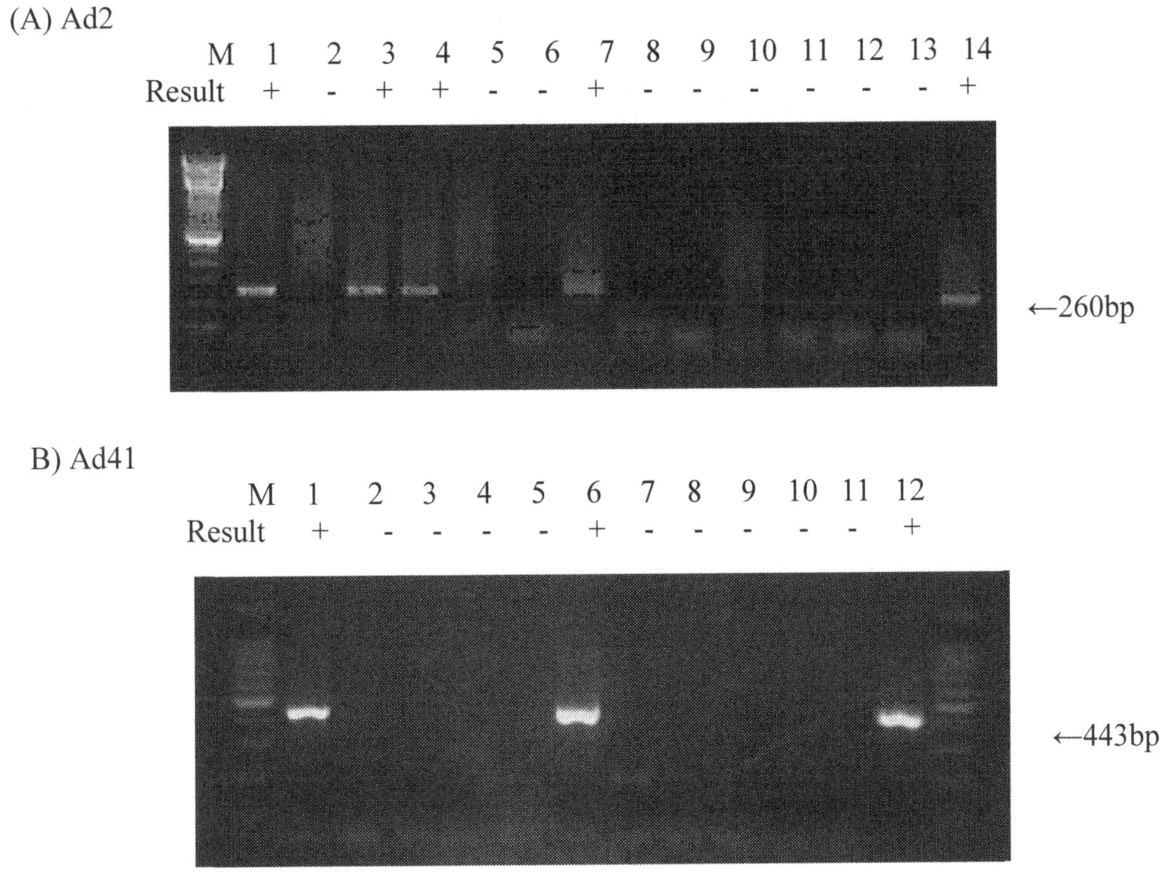

Figure 3.8. RT-PCR detection of Ad mRNA from cells inoculated with environmental water concentrates seeded with Ad. (A) lanes 1 and 2: RT-PCR of Ad2 mRNA from A549 cells infected with 20 and 2 IU in NC water concentrate; lane 3 and 4, RT-PCR of Ad2 mRNA from cells infected with 20 and 2 IU in GA water concentrate; lane 5, RT-PCR of mRNA in non-infected A549 cells; lane 6, negative control; lane 7, positive control; lane 8 to 9, PCR of Ad2 mRNA from A549 cells infected with 20, 2 IU in NC water concentrate; lane 10 to 11, PCR of Ad2 mRNA from cells infected with 20, 2 IU in GA water concentrate; lane 12, PCR of mRNA in non-infected A549 cells; lane 13, negative control; lane 14, positive control. (B) lane 1 to 3, RT-PCR of Ad 41 mRNA from G293 cells infected with 10, 1, 0.1 IU in GA water concentrate; lane 4, RT-PCR of mRNA in non-infected G293 cells; lane 5, negative control; lane 6, positive control; lane 7 to 9, PCR of Ad41 mRNA from G293 cells infected with 10, 1, 0.1 IU in GA water concentrate; lane 10, PCR of mRNA in non-infected G293 cells; lane 11, negative control; lane 12, positive control.

The results of this study indicate that Ad mRNA detection by RT-PCR is an effective, sensitive and rapid method to detect low levels of infectious Ad in water and clearly distinguish them from non-infectious viruses. Previous studies on the prevalence of Ad in surface and drinking water examined virus presence by DNA PCR. To overcome the potential for false-positive results in PCR due to the amplification of DNA from non-infectious viruses, we have developed a method to detect infectious Ad based on RT-PCR amplification of viral mRNA. Because only infectious viruses can produce mRNA, this assay is specific for infectious viruses. The findings of this study are particularly applicable for quantitative detection of infectious enteric Ad (Ad40 and Ad41), because these viruses do not produce clear CPE in cell cultures, even after prolonged incubation periods.

The results of this study also show that adenoviruses inactivated by low doses of free chlorine or high doses of UV can still be detected by PCR. Because they do not produce detectable mRNA in cell culture, they are not detected by the newly developed mRNA RT-PCR method. These results indicate that naked viral DNA or DNA fragments or DNA associated with inactivated virus particles are not detected by the mRNA RT-PCR method. This is because such DNA probably does not infect cells and produce viral mRNA. Because adenoviruses inactivated by both chemical and physical agents were not detected by mRNA RT-PCR, the developed method detects only infectious viruses and not those inactivated by disinfection processes.

Compared to Ad2, Ad41 is more difficult to culture, and the copy numbers of Ad41 RNAs are much lower than those of Ad2 (Tiemessen, Nel and Kidd, 1996). However, Ad41 has been successfully propagated in several cell lines including G293 cells, a line of human embryo kidney cells transformed with Ad5 E1 region (Graham et al. 1977; Tiemessen and Kidd 1996). The results of our study indicated that mRNA RT-PCR of Ad41 requires longer incubation times PI and has a lower sensitivity than that of Ad2 mRNA RT-PCR detection. However, using hemi-nested PCR, it was possible to detect mRNA from cells infected with only a few Ad41 as early as 3 days PI. Hemi-nested PCR can also increase specificity as well as sensitivity in viral nucleic acid detection because of the need to recognize in succession two different target sequences of nucleic acid.

Our data indicated a somewhat lower sensitivity of mRNA RT-PCR detection when Ad were suspended in environmental water sample concentrates as compared to PBS. There was a ten-fold sensitivity difference when inoculum was contained in two different environmental water concentrates (Figure 3.8). These data suggest that inhibitors either to virus infection in cell culture or interfering with virus recovery from inoculated cell cultures may be present in environmental water concentrates. Various particles, dissolved solutes and different conditions of ionic strength in water concentrates can potentially influence virus adsorption to particles, inhibit virus infection of cells, or reduce mRNA production in infected cells. Any of these effects could lead to lower sensitivity of mRNA RT-PCR for adenoviruses in inoculated cell cultures. Although sensitivity was lower, as few as 2-20 IU of Ad were detected.

CELL RECEPTOR CAPTURE and IMMUNOCAPTURE FOR RT-PCR DETECTION OF INFECTIOUS CVB3

sCAR production and evaluation

The sCAR protein produced and purified as described in Methods and Materials was evaluated by western blot and colloidal blue stain of total protein. Results are shown in Figure 3.9. The 36 kDa band is the correct size of the extracellular domain of sCAR as described by Roelvink et al (1998). This band is a doublet specifically identified by the use of the anti-FLAG antibody, which reacts with the FLAG tag on the sCAR protein produced in the insect cells infected with the baculovirus containing the code for the sCAR protein. The protein harvested from TN5-B14 cells inoculated with an MOI of 5 or 10 baculovirus units yielded similar results (Fig 3.9 columns A and B) as shown in the western blot analysis in the left column. When total protein is measured from the cells on a colloidal blue stain, the sCAR protein is not really visible in contrast to many other proteins in the crude preparation, right column of Figure 3.9.

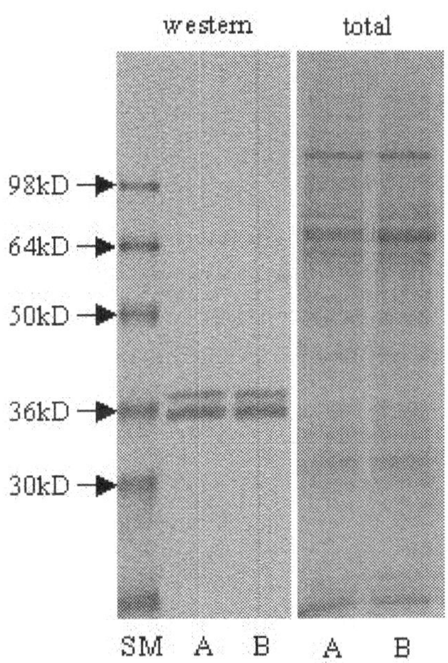

Figure 3.9. Comparison of western blot and total protein in cells 4 days PI at (A) 5 MOI and at (B) 10 MOI.

The purity of the sCAR product is demonstrated in Fig 3.10. Fractions of the purified protein from the affinity column were measured by the Bradford method to calculate yield of sCAR. Since three fractions contained purified protein, these were pooled for the final product. The yield was 150 µg scar purified sCAR/T150 flask, which is adequate for about 600 RT-PCR reactions. The results of these studies indicate that sCAR of excellent purity and yield was obtained. This purified material was used in subsequent experiments on recovery and RT-PCR detection of coxsackieviruses and adenoviruses.

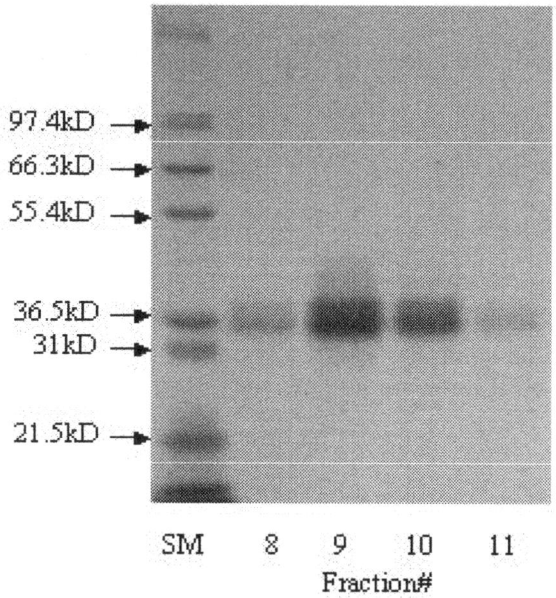

Figure 3.10. Analysis of sCAR purified with anti-FLAG affinity column. Fractions were collected from the column and analyzed by western blot. Supernatant solution of disrupted Tn5-B14 cells and culture broth were loaded on gels for column purification. Protein concentration of fraction 9 is 54 μg/ml, of fraction 10 is 53 μg/ml. After western blot, the membrane was stained with colloidal blue.

Neutralization of CVB3 by sCAR

The purified sCAR material described above was used to evaluate neutralization of CVB3 infectivity. If the sCAR is active and effective, it should react specifically with viruses by binding to the site that normally attaches to the CAR receptor on the surface of the host cell. However, when an infectious virus reacts with CAR in vitro and not on the surface of a live cell, the result of that binding reaction is to prevent subsequent virus infection of susceptible cells. Such a result is analogous to the loss of virus infectivity when the virus binds to an antibody molecule specific for or affected by the binding site on the virus surface that is responsible for attachment to a host cell. This loss of infectivity by specific binding to the virus surface is called neutralization. Results of a neutralization experiment using different concentrations of sCAR and a constant concentration of 100 infectious units of CVB3 are shown in Table 3.16. Ninety-five percent of CVB3 infectivity was neutralized by 0.175 μg of purified sCAR protein. This demonstrates that the binding to the soluble sCAR receptor protein neutralizes infectivity, and therefore, the sCAR is functional and active in binding of the virus at the site by which it binds to and infects host cells.

Table 3.16
Neutralization of CVB3 with different quantities of sCAR protein

CVB3 PFU	sCAR (µg)	Plate1	Plate2	Plate3	Mean (PFU)	SE (PFU)	Percent Neutralized
100	0.000	90	89	93	90.7	5.0	0.0
100	0.035	30	22	21	24.3	4.2	73.2
100	0.175	0	10	3	4.3	9.2	95.2
100	0.350	2	2	5	3.0	3.8	96.7
Negative Control		0	0	0	0.0		

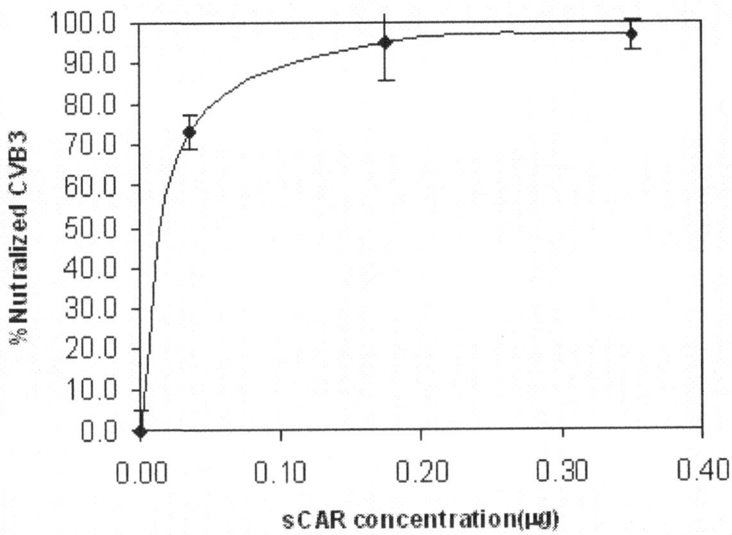

Figure 3.11. Graphic representation of neutralization of CVB3 by sCAR purified protein.

Table 3.17
Neutralization of Ad2 and Ad5 by sCAR

sCAR (μg)	Mean% Neutralized (Ad2)[a]	SE[b]	Mean % Neutralized (Ad5)[a]	SE[b]
0.035	29.7	3.5	6.5	10.1
0.175	45.8	4.6	20.5	1.8
0.350	59.8	17.0	63.0	15.6
2.000	63.8	4.3	82.7	11.5
5.000	95.0	10.0	96.2	3.8

[a] % neutralization is calculated based on (Co -Ct)x100/Co, where Co is the infectious unit count (PFU) without sCAR treatment and Ct is the infectious unit count with indicated amounts of sCAR.
[b] standard error

The neutralization of virus infectivity of sCAR was evaluated for Ad2 and Ad5 using a constant amount of virus (100 PFU) and different concentrations of sCAR ranging from 0.035 to 5.0 μg. Neutralization of Ad2 and Ad5 required 5 μg of sCAR purified protein in contrast to the 0.175 μg required for neutralization of CVB3. Several experiments were attempted using the sCAR-coated beads to capture Ad for detection by PCR, but they were unsuccessful. There are at least three factors contributing to the low efficiency of adenovirus capture by sCAR on beads for subsequent PCR detection: (1) the need for a high sCAR concentration to achieve efficient binding and neutralization, (2) the differences between coxsackieviruses and adenoviruses in their stearic association with sCAR, and (3) possible interference with Ad binding to sCAR due to excessive levels of virus binding site material in the virus preparation that is not actually associated with virus particles containing DNA, which acts as a competitive binder of sCAR. The extent to which excess non-virion virus binding material was responsible for the lack of virus binding was not investigated in this study and should be further explored in future studies by using purified virus suspensions freed of possible interfering non-virion sCAR binding material. The role of steric issues in Ad binding to sCAR also may be important. This is because the binding of Ad to the D1 extracellular functional region of sCAR is located on the other side of the structure from the CVB binding site (Nemerow 2000).

This different binding site and stereochemistry would make it very difficult or impossible to use sCAR to detect Ad with the current sCAR bead system. Perhaps a different type of surface with sCAR protein attached in a different way with a higher concentration of sCAR will be needed for efficient Ad capture for subsequent PCR. It was not possible to identify and develop such an alternative solid phase binding surface with attached sCAR within the resources of this project, and therefore, future studies would be required to address this need.

Ligand capture of CVB3 with sCAR on magnetic beads for RT-PCR detection

Amine-terminated beads coated with sCAR were used successfully to specifically capture CVB3. Binding of the sCAR to the bead surface is via the carboxy terminal end of the amine-terminated beads. Other coated beads used for capture of sCAR were less successful and yielded non-specific attachment of virus to the beads that could not be blocked with many different blocking agents tried (results not shown). Preliminary work demonstrated that the beads must be removed before the captured viruses are extracted for RNA and the extracted RNA subjected to RT-PCR. The method of heat release of RNA from the virus bound to beads did not yield consistent recovery of viral RNA.

Magnetic removal of the beads after capture of virus and then chemical lysis of captured viruses to release and purify viral RNA was necessary. The QIAamp viral kit was used to extract RNA. As described in Methods and Materials, after capture of the virus by sCAR coated beads, the virus containing material is removed when beads are separated using the magnet. After washes, the lysis buffer was applied to the beads, the beads removed, and the lysis buffer containing RNA transferred to a tube for continuation of the procedure of RNA extraction. The detection of sCAR with this method is shown in Figure 3.12. The sensitivity was 1 PFU of CVB3 in a reaction, and no non-specific viral RNA detection occurred when beads were not coated with sCAR but exposed to as much as 10^3 PFU/reaction of CVB3.

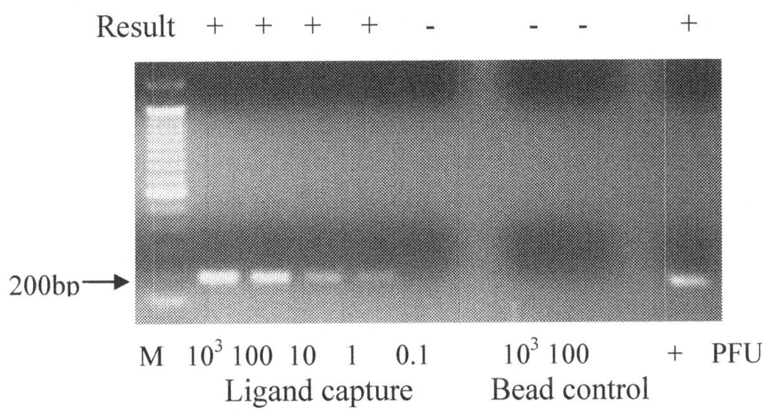

Figure 3.12. Ligand magnetic bead capture of CVB3 with purified sCAR on magnetic beads compared to control beads without sCAR

If the sCAR bead capture system is specific for intact, infectious viruses, it should not capture and detect viruses that have been inactivated with chemical disinfectants such as free chlorine. Therefore, CVB3 treated with 1 mg/L of free chlorine for 10 minutes (CT = 10 mg-min/L) and untreated CVB3 were tested for sCAR bead capture and RT-PCR detection in the same assay. The disinfected stock had no infectivity titer remaining when 0.2 mL aliquots were assayed in duplicate, and the same end-point RT-PCR titer as the control was obtained. CVB-3 and untreated CVB3 were tested in the same assay. Chlorine treated virus was not detected and the control virus reaction sensitivity was 1 PFU, as previously observed (Figure 3.13). This demonstrated that virus inactivated by the free chlorine disinfection method commonly used for drinking water treatment can not be detected in the assay.

Figure 3.13. RT-PCR detection of sCAR captured CVB3, control and chlorine treated. [a]Infectious unit (PFU) before disinfection. CVB3 treated with CT of 10 mg-min/L contained no infectivity.

Detection of wild-type virus and detection in water concentrates

There are six CVB types (types 1-6) that could be present in water and should be detected by the sCAR capture assay. To show that the sCAR capture method was applicable to other coxsackie B viruses, it was applied to a CVB4 that had been concentrated from sewage by PEG precipitation, followed by plaque purification and growth as described in detail in Chapter 2. This isolate was tested in a standard RT-PCR and in the sCAR capture assay (Figure 3.14 A and B). The RNA from as low as 0.1 PFU of virus was detected in the RT-PCR assay by direct extraction of RNA from the virus preparation (A). In contrast, the sCAR capture assay detected 10 PFU clearly and possibly 1 PFU (B) when viruses were captured on sCAR-coated beads and then processed for RNA extraction and RT-PCR amplification. These results indicate that wild-type viruses should be detected in the assay at relatively low levels (as little as 1 PFU) in the sCAR capture assay. These results demonstrate a greater sensitivity (lower limit of detectability) by the assay of viral RNA extracted directly from the virus preparation in comparison to the detection after the capture of intact virus particles with the sCAR bead system and subsequent RNA extraction. This difference in detectability by direct RT-PCR of the virus and after sCAR capture of functional and possibly infectious viruses could be a consequence of at least two factors: (1) the exclusion of viruses and viral RNA not associated with intact and functional virions, and/or (2) a lack of efficiency in virion capture and RNA extraction using the sCAR bead system. The discrimination between these two factors can not be made conclusively using only partially purified used viruses. To more conclusively address this issue, highly purified coxsackieviruses would need to be tested, and such an approach is recommended for future studies.

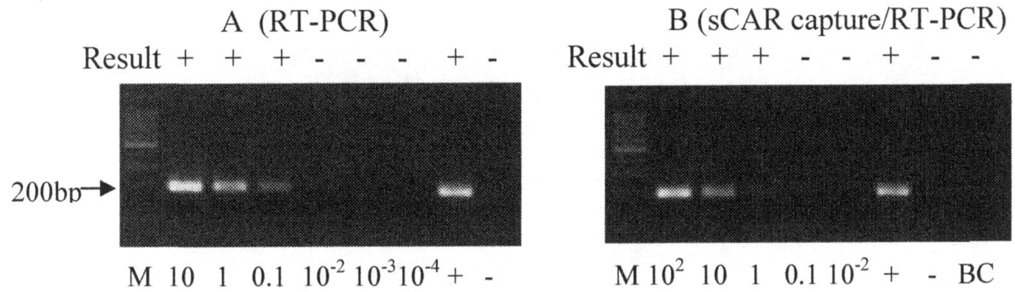

Figure 3.14. Detection of wild type CVB4 by standard RT-PCR analysis (A) and with the sCAR capture assay (B).

In order to determine if the sCAR bead capture system was effective and practical for detecting coxsackieviruses in field samples, its performance was evaluated for seeded field sample concentrates from surface water. Previous experiments were performed only with dilutions of virus in PBS. CVB3 was spiked into water sample concentrates prepared by PEG precipitation of liters of source water by the same methods as the utility collected samples and described in the Methods and Materials section of this report. The precipitate was resuspended in proportionally the same volumes as the samples from the utilities, which is 10 liters concentrated to 30 mL of water sample concentrate for seeding with viruses. The sCAR capture was then performed using 100 µl of water sample concentrate containing the indicated quantity of virus. Results in Figure 3.15 show that the lower detection limit of the laboratory strain CVB3 in surface water concentrates was 1 PFU, which is the same as previously observed in PBS diluent. No appreciable inhibition occurred in virus recovery and detection from the water sample concentrate, despite the likely presence of potential RT-PCR inhibitors in the concentrated pond water. This lack of inhibition in virus recovery and detection demonstrates that the system can perform well for the assay of actual field water samples.

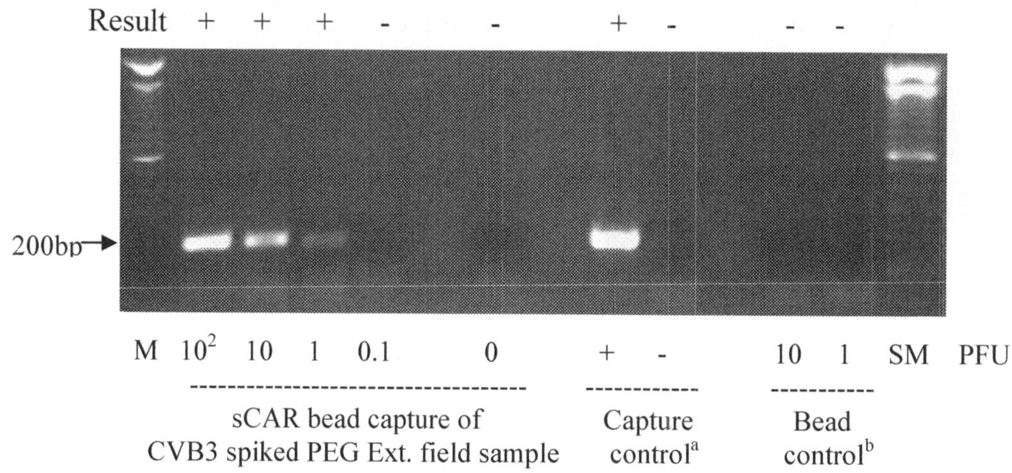

Figure 3.15. sCAR bead capture of CVB3 (laboratory propagated) spiked pond water sample concentrate

[a] Capture control: sCAR bead capture of 10 PFU of CVB3 (+) in PBS and PBS blank (-)

[b] Bead control: BSA coated bead (without sCAR) capture of CVB3 spiked in 100 μl of water sample concentrate.

A wild-type CVB3 isolate was obtained from a treated wastewater effluent by the culture of a wastewater effluent on BGM cells as described in Methods and Materials. Wastewater was concentrated by the same methods as the water samples. The cultured virus was identified by sequence analysis of a VP1 genomic region of 341 bases of nucleic acid sequence. The wild-type CVB3 is a Nancy strain, however, it has only 77% sequence homology with the laboratory Nancy strain in the genomic region and therefore is a distinct isolate. This CVB3 wild-type isolate was also analyzed for detection by direct analysis of RNA by RT-PCR and by sCAR capture followed by RT-PCR. Results are shown in Figure 3.16 A and B. The limit of detection with the direct RT-PCR was 0.1 PFU in 100 μl PBS (A) and 10 to 1 PFU in 100 μl of creek sample water concentrate (B). These results are similar to those observed with the CVB4 wild-type virus in PBS, and demonstrate that two separate infectious wild-type CVBs from wastewater sources can be detected in the sCAR capture RT-PCR assay.

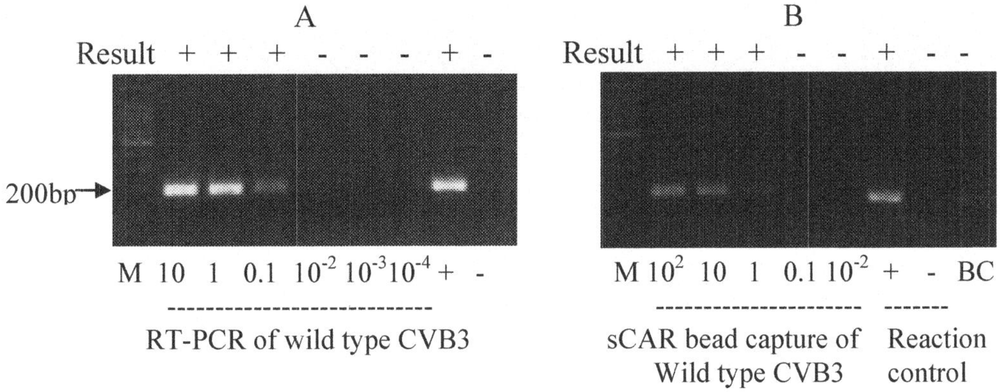

Figure 3.16. Detection of wild type CVB3 by standard RT-PCR analysis in PBS (A) and with the sCAR capture assay in spiked water sample concentrate (Bolin Creek) (B).

The wild-type CVB4 was also analyzed by the direct RT-PCR and RT-PCR after sCAR capture of virus diluted in a water sample concentrate. Results are shown in Figure 3.17. The concentrated virus preparation was diluted in a creek water sample concentrate prepared the same as river and pond concentrates. Results similar to those obtained with CVB3 wild-type virus diluted in water sample concentrate and similar to CVB4 diluted in PBS were obtained. By direct RT-PCR, 0.1 PFU was detected and after sCAR capture 10 to 1 PFU was detected.

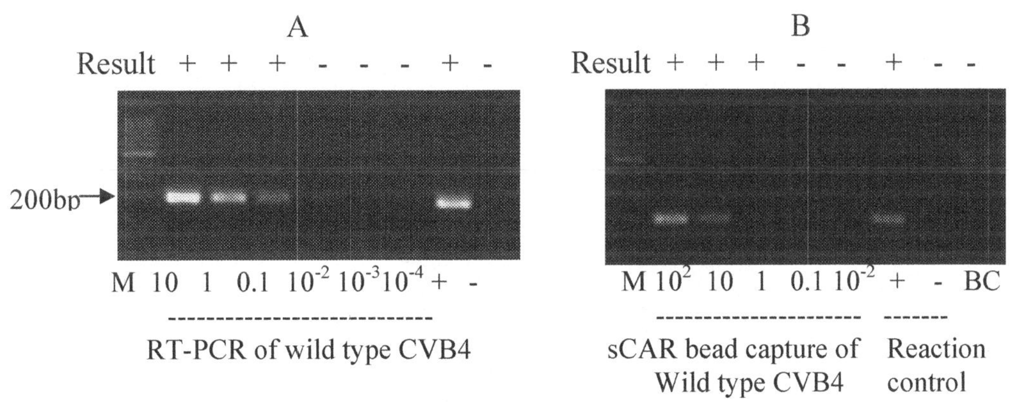

Figure3.17. Detection of wild type CVB4 by standard RT-PCR analysis in PBS (A) and with the sCAR capture assay in spiked water sample concentrate (B).

Capture of virus by sCAR-coated beads was evaluated in varying volumes of water sample concentrates from 50 μl to 1000 μl. Results are shown in Figure 3.18. Detection of CVB3 was at least 1 PFU and may have been lower than that based on the strength of the amplicon bands detected for sample volumes as large as 1000 μl. This is a ten-fold larger sample volume than used in the previous experiments. The use of the larger volumes will be useful in evaluation of increased volumes of water sample concentrates with the method and can increase the sensitivity.

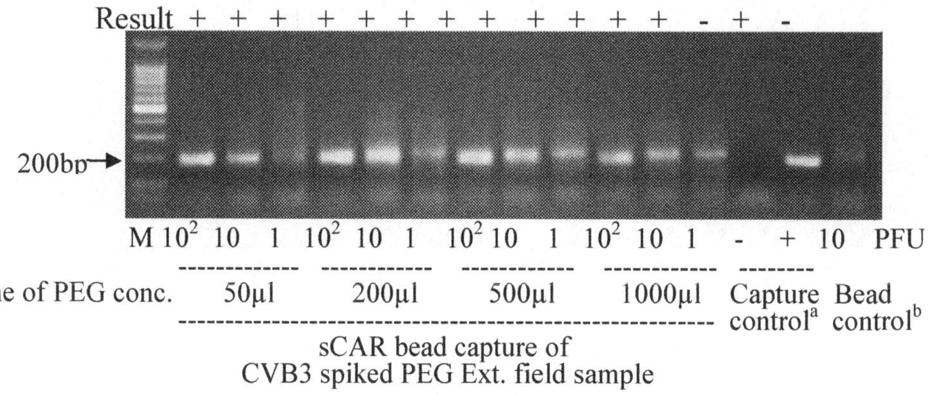

Figure 3.18. sCAR bead capture of CVB3 in spiked water sample concentrates (PEG conc.) of different volumes: 50 µl, 200 µl, 500µl, and 1000 µl
[a] Capture control: sCAR bead capture of CVB3 (+) in PBS and PBS blank (-)
[b] Bead control: BSA coated bead (without sCAR) capture of CVB3 spiked PEG

Magnetic bead immuno-capture and RT-PCR assay of CVB3

For capture and subsequent analysis of functional and potentially infectious viruses, an alternative approach besides using a functional cell receptor is the use of specific antibodies directed against virus surface epitopes for attachment to cells and associated with neutralization of virus infectivity. In this study a monoclonal, neutralizing antibody against coxsackievirus B3 was employed for virus capture prior to RNA extraction and RT-PCR amplification. Capture of CVB3 with magnetic beads coated with monoclonal antibody was evaluated for comparison with other assays. By direct RT-PCR, as low as 0.1 PFU was detected (Figure 3.19).

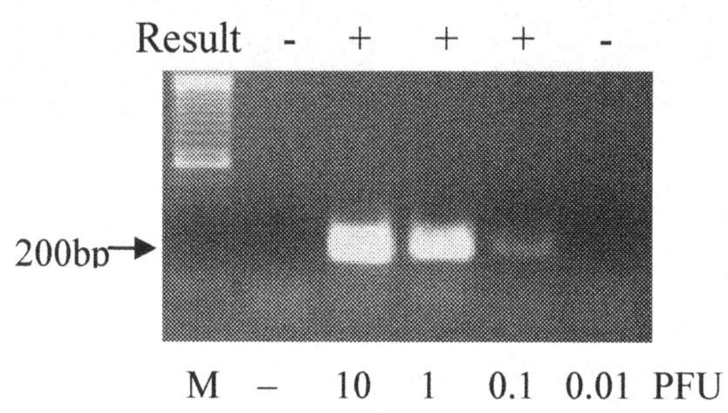

Figure 3.19. Direct RT-PCR of CVB3.

In contrast to detection by direct RT-PCR, the sensitivity of immunocapture followed by RT-PCR was 1 PFU (Figure 3.20). The 10-fold difference could indicate that only intact reactive virus particles are detected in the immunocapture RT-PCR assay system. If this is the case, the results suggest that a population of RNA containing particles or RNA not associated with virus particles at all, are not detected by the immunocapture assay. Detection of RNA in

RT-PCR with no prior step of immunocapture was higher if any particle containing RNA or any free viral RNA in the preparation was available for detection in the assay. In contrast, only those viruses capable of reacting with the antibody would be detected by the immunocapture plus RT-PCR assay system. As previously noted for sCAR captured viruses, this observed difference in detectability by direct RT-PCR of the virus and after immunocapture of functional and possibly infectious viruses could be a consequence of at least two factors: (1) the exclusion of viruses and viral RNA not associated with intact and functional (capturable) virions, and/or (2) a lack of efficiency in virion capture and RNA extraction using the immunocapture system. The use of highly purified viruses would be needed in this type of experiment in order to conclusively distinguish these two effects.

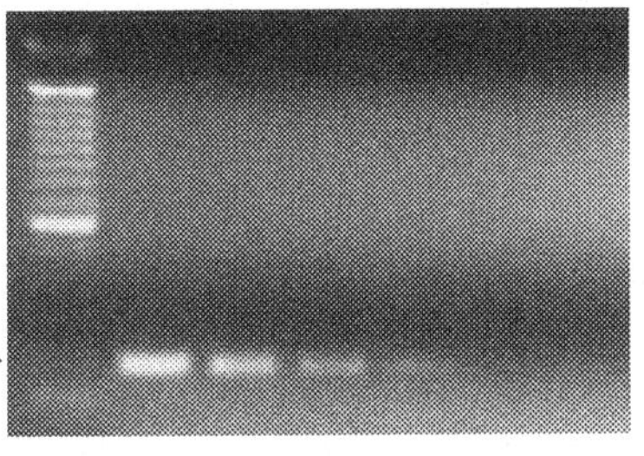

Figure 3.20. Immuno-magnetic bead capture of CVB3 with anti-CVB3-Mab coated beads

In summary, the results of these studies demonstrated that purified sCAR can be produced in sufficient quantity and quality for covalent attachment to magnetic beads to capture coxsackie B viruses. A lab-adapted strain of CVB3 and wild type CVB4 and CVB3 could be bound specifically and efficiently by the sCAR-coated beads for detection by standard RT-PCR after extraction of RNA by a column method. The lower detection limit was 1PFU per reaction on the magnetic sCAR coated beads. Capture of CVB3 with monoclonal antibody-coated immunobeads for subsequent RT-PCR detection also was effective. In contrast, the analysis of the same quantity of virus by extraction of the RNA directly without sCAR or monoclonal antibody bead capture yielded a lower detection limit of 0.1 to 0.01 PFU. The difference in RT-PCR detection between direct analysis and the sCAR or the immuno-capture with monoclonal antibody could indicate that a population of the virus with non-reactive capsids or the presence of viral RNA not associated with capsids is not detected in these assays. This is supported by the fact that chlorine inactivated virus was not detected in the assay. On the other hand, an alternative explanation for the reduced detection by the sCAR or monoclonal antibody bead capture method is reduced efficiency in either virion capture, RNA extraction, RT-PCR amplification or a combination of these steps.

This newly developed method can detect potentially infectious CVB within 12 hours or less. This method is promising for a specific and rapid approach to the detection of potentially infectious viruses in source, drinking and other environmental water. More work is needed, however, to better understand the mechanisms of virion capture and to examine the alternative outcomes of viral nucleic acid recovery and detection from virus preparations that have been disinfection by different mechanisms compared to infectious viruses. In addition, further development is need for sCAR bead capture of adenoviruses, which was not possible using the beads developed in this study.

REAL-TIME RT-PCR FOR DETECTION OF VIRUSES

Rapid Detection of Hepatitis A virus and Enterovirus in Water Sample Concentrates by Real-time RT-PCR

Evaluation of primers and probe for the detection of HAV and enteroviruses

The primers and probes described in Methods and Materials were tested on several HAV isolates to evaluate the application of the real-time RT-PCR for identification of human HAV genotypes. Representative strains of Genotype (or subgenotypes) 1A, 1B, II, III, and VII HAVs were positive with the primer/probe combination (Figure 3.21). A range of cycle numbers (also called crossing point or Ct) on the X-axis was observed for the samples. Generally fluorescence should appear by cycle 40 in a positive sample. Fluorescence was first observed from cycle 18 to 24 in these samples. Generally a 3.3 cycle differential represents one log difference in the number of nucleic acid copies between samples. This is only true, however, in ideal circumstances with no inhibitors of the reaction and without any mismatch of the probe. Only two of these samples have infectivity titers: 1B is HM-175, clone 24A,a laboratory strain with a cell culture infectivity titer, and 1A is HLD-2 (described in Methods and Materials) with known primate infectivity. The other samples are extracts of stool or liver homogenates with unknown infectivity titers. This assay demonstrates that representatives of these five genotypes of HAV can be detected by the primers and probe. In contrast, all enteroviruses tested including CVA8, CVA10, CVB3, CVB4, Echovirus 1, 27, and 29 and poliovirus type 1 were negative in the assay (data not shown).

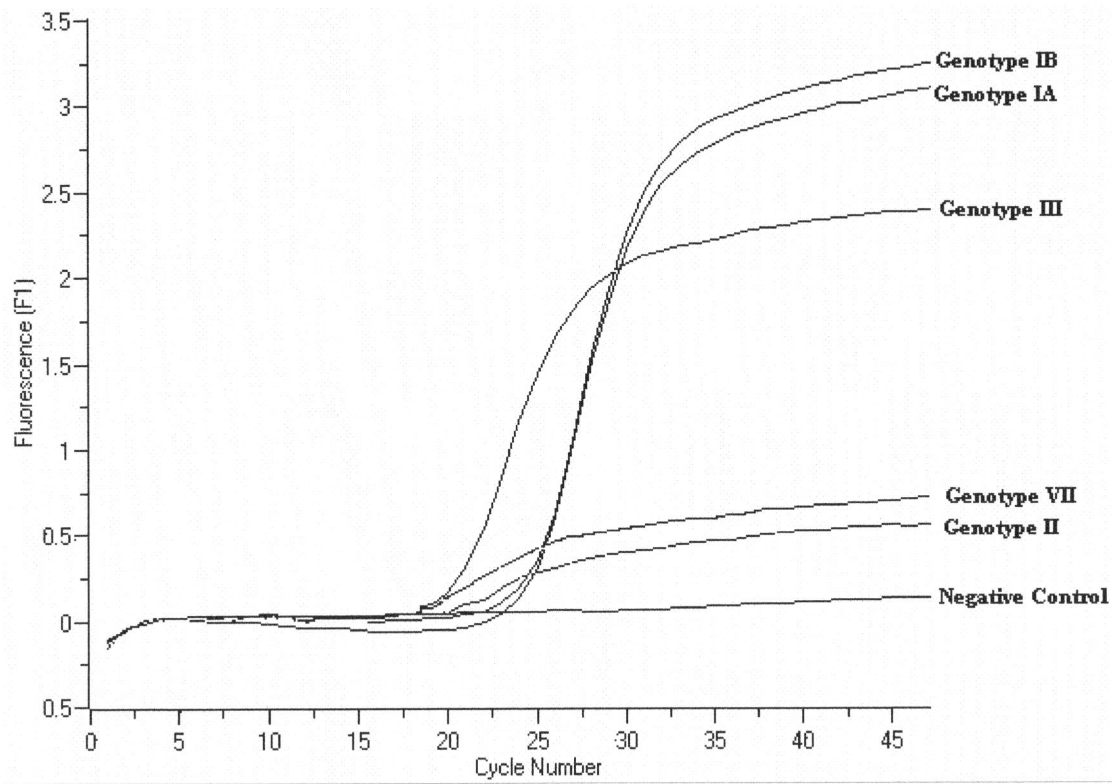

Figure 3.21. Detection of different genotypes of human HAV.

Primers amplifying a 145 bp fragment in the 5'-non-coding region (5'NCR) of enteroviruses and an internal TaqMan probe were specific for a panel of enterovirus isolates, including CVB3, CVA8, CVA10, Echovirus 27 and 29, and poliovirus 1 (Figure 3.22). Fluorescence was detectable for all enteroviruses shown at 20-22 cycles indicating good specificity. Echovirus 1 and CVB4 were also positive (data not included in this graph).

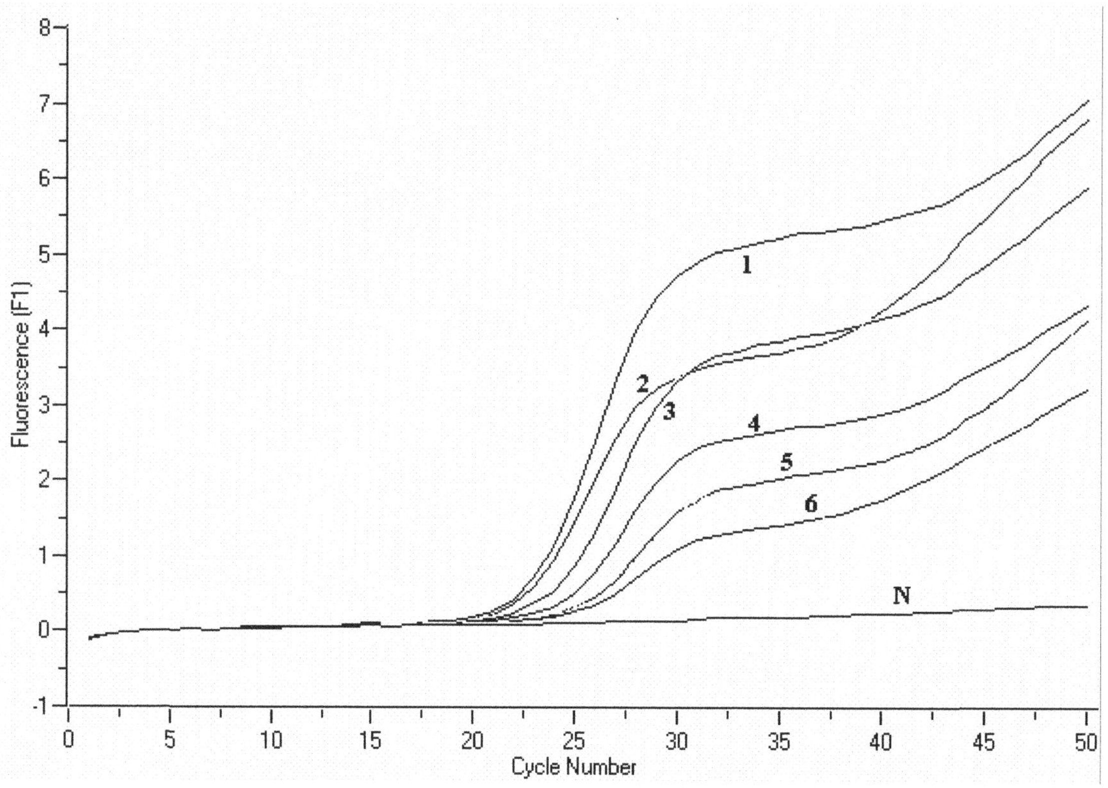

Figure 3.22. Specificity of the enterovirus TaqMan assay for detection of various isolates of enterovirus: (1) Echovirus 27, (2) CVB3, (3) CVA8, (4) Polio 1 (5) Echovirus 29, (6) CVA10, and N, Negative.

Duplex real-time RT-PCR for simultaneous detection of HAV and CVB3

Primer pair concentrations were optimized separately for each agent (HAV and CVB3 as a model enterovirus) and for simultaneous detection of HAV and CVB3 in a duplex RT-PCR. Each set of primers amplified only one virus type and no cross-reactivity was found between the individual primers and non-target virus during monoplex RT-PCR. HAV primers and probe were demonstrated to detect all human genotypes in the previous figure (Figure 3.21). Enterovirus primers and probe detect a broad range of viruses based on comparisons in GenBank, furthermore detection of several enterovirus types is shown in Figure 3.22.

To determine if multiplex RT-PCR amplified only the target virus (es) and the correct amplicon size was obtained, the amplified material was subjected to agarose gel electrophoresis (Figure 3.23). This further confirmed that each fragment found in multiplex RT-PCR was amplified from one type of virus. The correct size amplicon for HAV (89 bp) and CVB3 (145 bp), lanes 1 and 3 respectively, was seen when each was assayed separately. In the duplex assay both amplicons are present (lane 5, Figure 3.20).

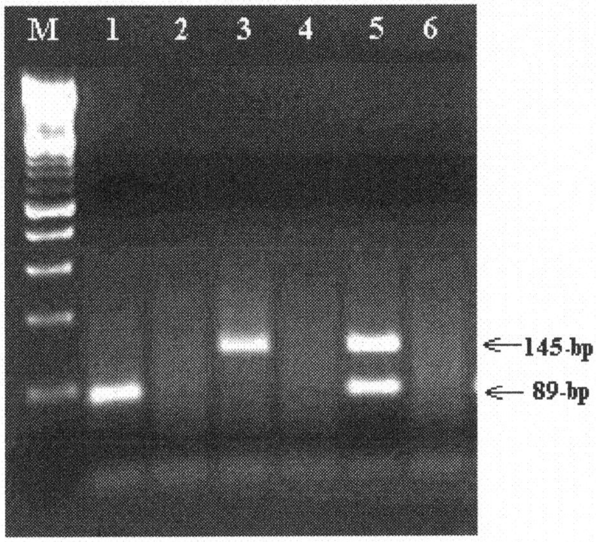

Figure 3.23. Agarose gel electrophoresis of PCR products from HAV and CVB3 real-time RT-PCR using primers designed to amplify the 89 (HAV amplicon) and 145-bp (CVB3 amplicon). Lanes: M, 100-bp molecular size standards; 1, HAV; 2, Negative control; 3, CVB3; 4, Negative control; 5, HAV and CVB3 template RNA; 6, Negative control for lane 5. Arrows indicate the relative positions of the amplified fragments.

When the mixed TaqMan probe was used in multiplex format the probe reacted with only one type of virus, and no cross-reactivity was found between the individual probe and the non-target RT-PCR amplicons during real-time TaqMan RT-PCR (Table 3.18). This table shows the R^2 values and standard curve slopes calculated by the instrument for each set of experimental targets. The R^2 value (always between 0 and 1) is an indication of the quality of the fit of the standard curve to the standard data points plotted, with values closer to 1 indicating a better fit of the data to the line. The slope of the standard curve is directly related to the average efficiency of amplification throughout the cycling program and may be used to calculate the PCR efficiency for a given template in a given experiment. In turn, the efficiency gives valuable information about the chemistry of the real-time RT-PCR reaction. A slope of -3.322 represents an efficiency of 100%. The slope of the standard curve is related to the efficiency through the following equation:

$\text{Efficiency} = [10^{(-1/\text{slope})}]-1.$

The data shown in Table 3.18 demonstrates by using different labeled probes for each virus target and only reading fluorescence for one probe of the selected virus target that the efficiency of the reaction is maintained when both probes are present.

Table 3.18
Statistical data on multiplex TaqMan RT-PCR quantification in the LightCycler

Assay	Linear regression curve (Y= -bc + a)	Correlation coefficient (R^2)	Range (PFU)
HAV Multiplex TaqMan[a]	-3.17c+40.70	1.00	0.4-4000
CVB3 Multiplex TaqMan[b]	-3.57c+40.74	1.00	0.4-4000

[a] Primer/probes were added for both HAV (FAM-Probe) and CVB3 (ROX probe) but only recorded for the FAM value (HAV)
[b]Primer/probes were added for both HAV (ROX-Probe) and CVB3 (FAM probe) but only recorded for FAM value (CVB3).

Each double stranded DNA product has its own specific melting temperature (T_m), which is defined as the temperature at which 50% of the DNA becomes single stranded, and 50% remains double stranded. Length and GC content are the most important factors determining T_m. Since SYBR green I dye binds within the minor groove of dsDNA, the fluorescent signal steadily decreases as the DNA melts. This increase and decrease in binding can be measured on the LightCycler during the real-time RT-PCR. Determining the T_m of PCR products provides confirmation about the PCR product identity and differentiation of specific PCR products from non-specific products, such as primer dimers.

In the following example, a typical melting curve analysis is described. The peaks created by any primer dimers, due to their shorter length, melt at a lower temperature, whereas the specific amplification product has a higher T_m. In the present study primer dimers were not observed, consequently there was no peak for primer dimers. Melting curve analysis of SYBR Green RT-PCR differentiated between Enterovirus (T_m 84 °C) and HAV (T_m 79 °C) amplicons, which is shown in Figure 3.24. In initial experiments primers were tested for possible interaction to make sure that they could be used together in multiplex format. In reaction 1 (CVB3) the correct T_m of 84 °C is observed, in reaction 2 (HAV) it is 79 °C, and when both are combined in one assay, peaks at both T_ms can be seen (3) in Figure 3.24.

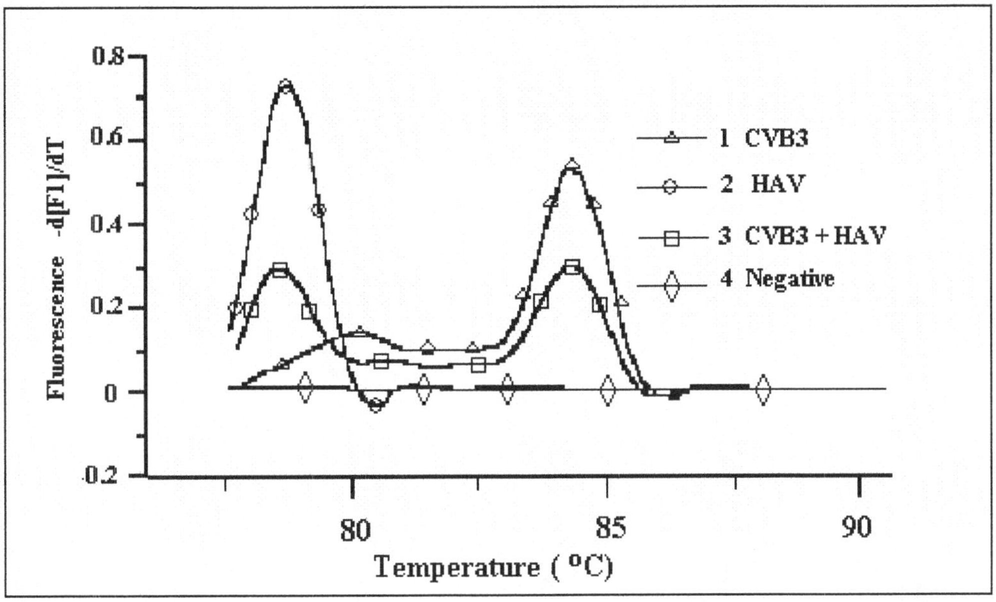

Figure 3.24. Multiplex SYBR Green RT-PCR. Tm: 84 ^{0}C for CVB3 (1) and 79 ^{0}C for HAV (2), Sample 3 contains CVB3 and HAV.

The sensitivity of the duplex TaqMan RT-PCR assay for HAV and for CVB3 was determined using samples with known infectivity titers. Ten-fold, serially diluted samples of HAV and CVB3 were amplified using the multiplex TaqMan assay. Detection of as low as 0.8 PFU of HAV and 0.8 PFU of CVB3 was observed (Figure 3.25). The cycle number of appearance of product as measured by fluorescence was about 33- 35 cycles for each of the viruses when 0.8 PFU was included in the reaction. With increasing quantity of virus in the reactions, the fluorescence appeared at decreasing cycle number, to as low as 20-22 with 8000 PFU present. In some experiments as few as 0.08 PFU was detectable within 40 cycles, but this was not consistent (data not shown).

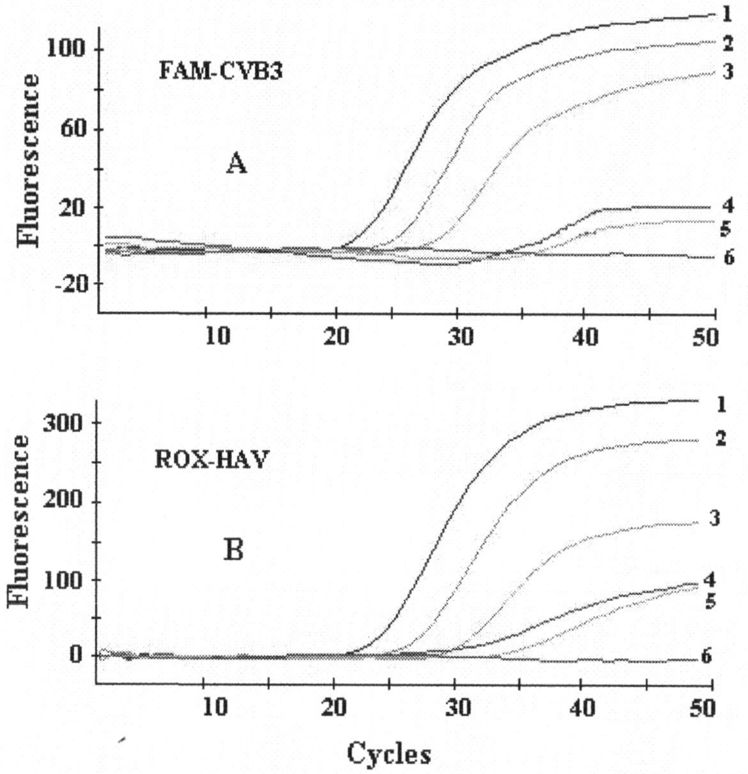

Figure 3.25. Multiplex RT-PCR for simultaneous detection of CVB3 (A) and HAV (B) at different dilutions: (1) 8000 PFU, (2) 800 PFU, (3) 80 PFU, (4) 8 PFU, (5) 0.8 PFU and (6) Negative control

Evaluation of the methods for detection of viruses in environmental samples

To demonstrate the practical application of HAV and enterovirus real-time RT-PCR, spiked water sample concentrates were examined for both HAV and CVB3. Source water samples from two different locations concentrated as described in materials were spiked with defined quantities of HAV and CVB3. RNA was extracted with the QIAamp viral kit lysis buffer (Qiagen, Valencia, CA) or the GuSCN lysis buffer, and both were applied to the Qiagen mini columns for RNA purification. Recovery of CVB3 and HAV spiked in water concentrates was evaluated for a range of 10-fold serial dilutions for the virus, ranging from 800 to 0.8 PFU (Table 3.19). Good correlation between viral load (800-0.8 PFU) and the crossing point or cycle number was observed with both lysis buffers. Therefore, the GuSCN method was used in subsequent experiments. With the GuSCN lysis buffer, crossing points in a spiked water concentrate, containing comparable viral loads were linear over 4 orders of magnitude, with R^2 values being greater than 0.97 for both HAV and CVB3 (Table 3.19). These experiments demonstrate that enteric viruses can be detected by real-time RT-PCR at very low levels in concentrated river and pond water. Results shown here are from experiments with river water sample concentrates. Experiments were performed three times with similar results, two times with separate sample concentrates of river water, and a third with a pond water sample concentrate.

Table 3.19.
Comparative sensitivity of Qiagen lysis buffer and GuSCN lysis buffer

Extraction	PFU per reaction						
	800	80	8	0.8	Slope	Intercept	R^2
QIAamp viral lysis buffer A.CVB3 B. HAV	 26.06[1] 26.99	 28.88 30.55	 33.31 37.98	 35.65 37.98	 -3.321 -3.670	 32.63 37.96	 0.99 1.00
GuSCN lysis buffer A.CVB3 B. HAV	 24.03 23.58	 28.36 26.71	 31.70 34.19	 33.97 34.19	 -3.317 -3.56	 34.49 34.05	 0.99 1.00

[1] Crossing point for each parameter

Parallel recovery trials with HAV and CVB3 demonstrated that recovery efficiency in water sample concentrates was equal to that in PBS. The RNA extraction protocol was performed on a midi column using extraction of a 4 mL spiked water sample concentrate with the newly developed GuSCN buffer described in Chapter 2. A range of concentrations of HAV and CVB3 were spiked into water samples for extraction, followed by analysis by TaqMan RT-PCR (Table 3.20). The correlation coefficient was the same for both viruses in water concentrate or PBS at a range of 80-8000 PFU, indicating efficient reactions (as described above for Table 3.18). When the RNA extraction protocol with a midi column was used for 4 mL of inoculated water sample concentrate, there was no loss of the sensitivity of detection. As low as 8 PFU inoculated in the water sample concentrate could be detected. The results did not follow a standard log-linear (exponential) curve over 5 orders of magnitude (0.8-8000); however, there was a log-linear curve over three orders of magnitude (80-8000). The lack of log-linearity of the curve for RT-PCR amplification suggested that non-specific inhibitors interfered in the reaction.

Table 3.20
Statistical data on TaqMan RT-PCR quantification in the LightCycler

Assay	Linear regression curve (Y= -bc + a)	Correlation coefficient (R^2)	Range (PFU)
HAV TaqMan[a]	-3.12c+35.00	1.00	80-8000
CVB3 TaqMan[a]	-3.99c+35.15	0.99	80-8000
HAV TaqMan[b]	-3.40c+38.07	1.00	80-8000
CVB3 TaqMan[b]	-3.65c+36.01	0.97	80-8000

[a] RNA extracted by GuSCN lysis buffer from PBS containing the indicated PFU of CVB3 and HAV

[b] RNA extracted by GuSCN lysis buffer from a water sample concentrate spiked with the indicated PFU of CVB3 and HAV.

RNA precipitation for increased sensitivity of detection by RT-PCR

The recovery sensitivity or the lower limit of virus detection by real time quantitative RT-PCR can be improved even more after the use of column purification and concentration of nucleic acids by adding an RNA precipitation step for further RNA concentration. All nucleic acid from one 4-mL extraction can potentially be evaluated in one reaction. Data presented in Table 3.21 demonstrates that excellent recovery of RNA post column can be obtained with several alcohol precipitation methods. As low as 0.8 PFU/reaction of HAV and of CVB3 could be detected with all methods. The Ct values were 35 – 36 for all methods when 0.8 PFU was evaluated. The control sample was RNA of equal volume evaluated before precipitation. The glycoblue method was chosen for use since it gives equally good recovery and is easy to visualize.

Table 3.21
Comparative sensitivity of different RNA reconcentration methods

Extraction	PFU/reaction						
	800	80	8	0.8	Slope	Intercept	R^2
CVB3							
A. Control	25.77[1]	29.50	33.31	35.49	-3.298	32.67	0.99
B. EtOH (Glycogen)	25.37	28.44	32.28	35.97	-3.565	32.30	1.0
C. EtOH (Glycoblue)	25.39	29.25	32.86	34.19	-3.001	31.92	0.98
D. Isopropanol	24.56	27.78	30.93	34.14	-3.187	34.13	1.0
HAV							
A. Control	25.54	28.67	31.74	35.06	-3.165	35.0	1.0
B. EtOH (Glycogen)	26.02	30.48	34.47	35.56	-3.260	36.53	0.97
C. EtOH (Glycoblue)	25.93	29.72	32.45	35.75	-3.221	35.79	1.0
D. Isopropanol	25.89	30.10	33.53	36.03	-3.417	36.53	0.99

[1]Crossing point value for each parameter

The Glycoblue RNA precipitation method described above (Method C) was chosen for precipitation of the RNA following the extraction on midi columns of 4.0-mL volumes of water sample concentrates. The modeling experiments presented in this study with CVB3 and HAV demonstrate the sensitivity and specificity of the recovery method developed, which includes extraction of 4.0 mL of water sample concentrate on the midi column, followed by RNA precipitation. This combined approach permitted examination of 8% of the total volume of the water sample concentrate for each type of virus, which corresponds to a water volume of 1.13 liters.

Rapid detection of Noroviruses Genogroup I with a TaqMan RT-PCR

Traditionally nucleotide variations found in the RNA polymerase gene within ORF1 have provided classification of Noroviruses into two main genetic groups, genogroups 1 and 2. Two oligonucleotides were designed in this region to amplify 96 bases of target, and a genogroup I-specific oligonucleotide fluorescent TaqMan probe also was designed for use in real-time RT-PCR. The degenerate forward primer corresponded to conserved regions of the RNA polymerase gene and when used with the non-degenerate reverse primer, they amplified a 96-base-pair (bp) fragment. Amplified sequences were detected by a TaqMan probe labeled with fluorescent reporter dye and quencher dye located internal to the priming sites.

When these reagents were tested in clinical samples, the TaqMan RT-PCR assay successfully achieved automated PCR amplification, detection and analysis of noroviruses. Data presented in Table 3.22 demonstrate that the real-time, quantitative RT-PCR assay is specific for genogroup I NV and that all clusters can be detected by the assay. Twenty-two samples were correctly identified with a single synthetic oligonucleotide primer pair to successfully amplify all of the genotype I viruses, including Desert shield, Southampton, Norwalk and Chiba in stool samples. There was no amplification of these target regions detectable in any of 17 genotype II specimens, indicating complete specificity in detection of genotype I. However, fluorescence signals differed greatly between samples and these differences could be due to the minor sequence variation between the primers and the target cDNAs of the individual target viral genomes. The method has the potential to be applied to real-time, quantitative Norovirus detection in environmental samples.

Table 3.22
Comparison of TaqMan based RT-PCR with Conventional RT-PCR for detection of different clusters of Genogroup I Noroviruses

RNA Extraction Protocol	No. of samples tested	No. of Positive Samples	
		TaqMan RT-PCR	Agarose Gel
<u>Nuclisens</u>			
Norwalk (GI.1)	2	2	2
Southampton (GI.2)	2	2	2
Desert Shield (GI.3)	2	2	2
Stavanger (GI.3b)	2	2	2
Chiba (GI.4)	2	2	1
Musgrove (GI.5)	NT	NT	NT
Hesse (GI.6)	1	1	1
Winchester (GI.7)	NT	NT	NT
<u>Manual</u>			
Norwalk (GI.1)	1	1	1
Southampton (GI.2)	2	2	2
Desert Shield (GI.3)	3	3	3
Stavanger (GI.3b)	1	1	1
Chiba (GI.4)	1	1	1
Musgrove (GI.5)	1	1	1
Hesse (GI.6)	1	1	1
Winchester (GI.7)	1	1	1

NT-Not tested; clusters are named based on CDC nomenclature

There were no inhibitory effects in real-time, quantitative RT-PCR amplification using RNA extracted with the NucliSens RNA automatic extraction system. This was determined by spiking HAV and CVB3 virus in stool samples followed by NucliSens extraction and Taqman detection (data not shown). The Nuclisens RNA extraction system is based on the Boom method (Boom et al., 1990) and is described in Chapter 2 Methods and Materials.

In summary, the real-time RT-PCR assay described here targets all clusters of genogroup I Noroviruses and will be useful for direct detection of these viruses from fecal and environmental samples in less than 2 h. Because the real-time RT-PCR can detect as low as 40 copies of genome equivalent (data not shown), it should provide rapid and sensitive detection in clinical/environmental samples containing low copy number of genogroup I Norovirus. The proposed TaqMan RT-PCR for detection of genogroup I can potentially be coupled with another TaqMan RT-PCR for detection of genogroup II simultaneously in a multiplex format as we have described for enteroviruses and HAV. Due to limitations of time and other resources, we were not able to develop and evaluate a genotype II real-time RT-PCR assay in this study.

Detection and Quantification of Adenovirus species C, D, and F with Real-Time PCR

Evaluation of primers and probes for detection of Ad species

A real-time, quantitative PCR assay was developed for the detection of Ad. The primer pair and TaqMan probe were designed for detection of isolates in species C (including Ad 1, 2, 5, and 6), D (Ad 8, 19, and others), and F (Ad 40 and Ad 41) with primers specific for the hexon region. The results for detection of the seven Ad serotypes representing species C (serotypes 1,2,5 and 6), D (serotype 8) and F (serotype 40 and 41), in the real-time PCR assay are shown in Figure 3.26. Amplification of Ad 2 and Ad 5 was apparent at cycle number (or crossing point) 12, whereas Ad 40 and 41 were detectable at cycle 30 and 20 respectively. The increased fluorescence seen with Ad 40 and 41 is due to the perfect match of the probe for these types.

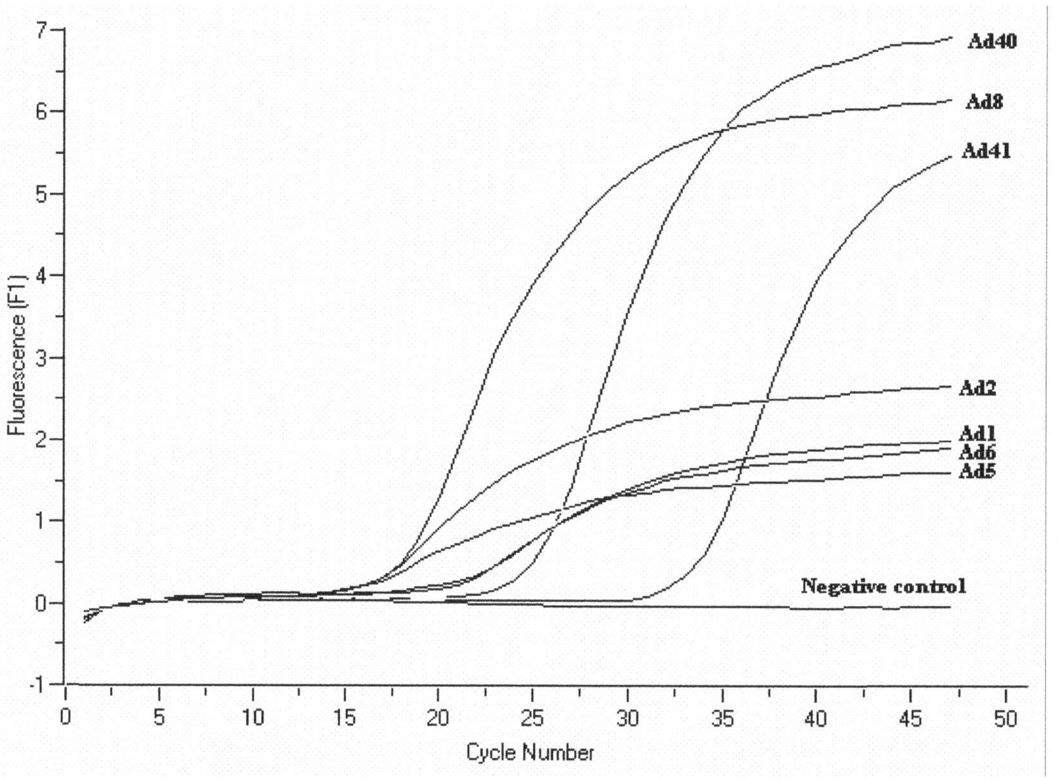

Figure 3.26. Typical amplification curve of species C (Ad 1, 2, 5,and 6), D (Ad 8) and F (Ad 40 and 41) from a panel of isolates with the hexon TaqMan PCR.

Another primer and probe set was developed to detect only species F adenoviruses (Ad 40 and 41). Since Ad 40 and 41 are the most common enteric Ad, detection of the F species including these subtypes will be useful for detection by real-time PCR in water samples. The primer set and probe reagents for 40 and 41 were tested on Ad 40 and 41 and results are shown in Figure 3.27. Early rise in fluorescence is seen at cycles 20 and 28 with Ad41 and Ad40 indicating good detection of these viruses.

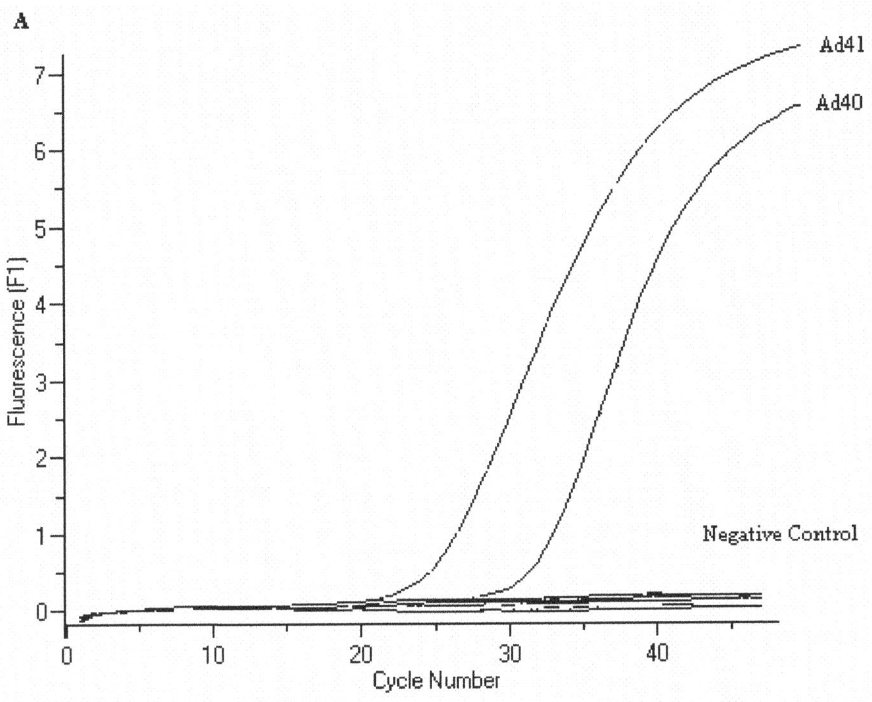

Figure 3.27. Detection of Ad40 and Ad41 by the fiber-gene based F subtype specific TaqMan PCR.

Detection of Ad2 and Ad41 with the hexon TaqMan PCR over a range of 7 $\log_{10}$ to 4 $\log_{10}$ respectively is shown in Figure 3.28 A and B. The infectious units (IU) measured as described in Methods and Materials was used to calculate the detection sensitivity. The lower limit of detection of Ad2 (Figure 3.28 A) was 0.1 PFU, which was detectable at about 38 cycles (within the maximum cycle number for detection of 40). Ad 41 (Figure 3.28B) was also detectable within 40 cycles with a sensitivity of 0.01 IU based on the infectivity analysis. The TaqMan assay sensitivity was comparable to that of the nested PCR when performed according to previously reported procedures (Xu et al. 2000) (data not shown) The TaqMan probe contains three mismatches in the 3' region when compared with sequences of the hexon genes of Ad2 and Ad5. However, these mismatches in the TaqMan probe did not alter the sensitivity or specificity of the assay for the detection of representative serotypes from species C (Ad 2 and 5) and F (Ad 40 and 41). Although differences in total fluorescence were seen in Figure 3.26 between Ad 2 and Ad 41 reactions, data in Figure 3.28 indicates that the sensitivity of detection is similar for the two different viruses. The limit of detection as measured by IU is only 10 fold different (0.1 IU for Ad2 and 0.01 for Ad 41) in Figure 3.28, which may be explained by the greater number of DNA containing particles in Ad 41. The particle to infectivity ratio of Ad 40 produced in cell culture is up to 1000 fold higher than for Ad2 (Mautner, 1999). We have not performed a particle count of our virus stocks.

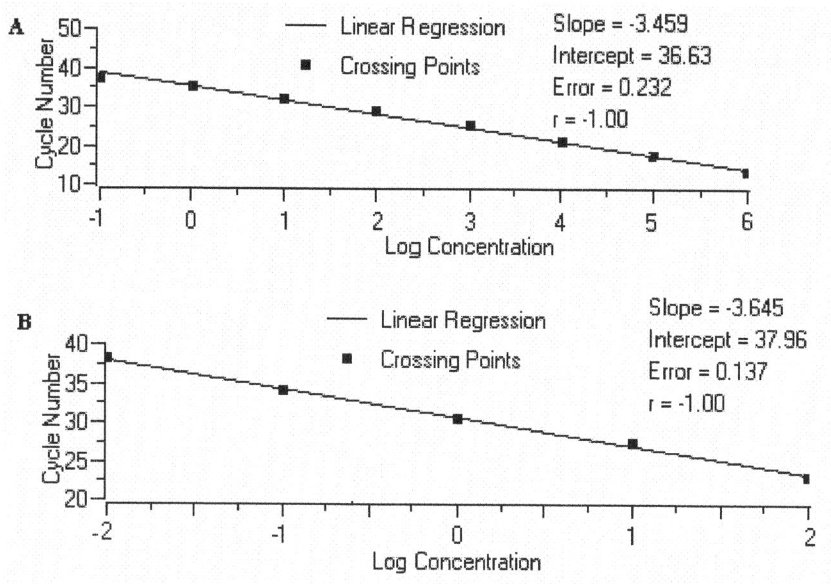

Figure 3.28. Standard curve of hexon based TaqMan PCR assay for Ad. Results are presented as the threshold cycle (Ct) values of serially diluted Ad2 (A), and Ad41 (B) with virus concentrations expressed as infectivity (IU or PFU)

Detection of Ad 41 with the fiber gene TaqMan PCR over a range of 4 log_{10} is shown in Figure 3.29. Sensitivity of detection of Ad41 with the fiber gene assay is also 0.01 IU, obtained at 33-34 cycles.

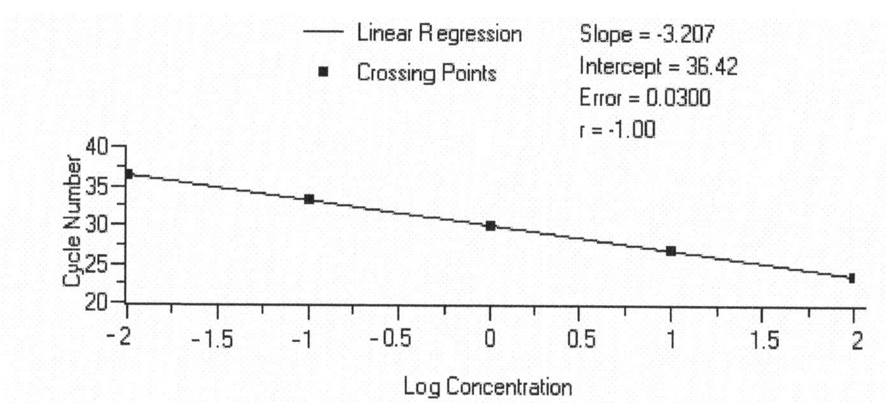

Figure 3.29. Standard curve of fiber-gene based TaqMan PCR assay, the threshold cycle (Ct) values of serially diluted Ad41. The standard curve was expressed as Ct vs. virus IU per assay (2 ul/assay).

Detection of Ad from clinical specimens, validation of primers and probes

A total of 44 clinical isolates of adenovirus species A to F were tested using the newly developed TaqMan, real-time RT-PCR hexon and fiber gene assays for Ad. The detection of Ad types belonging to different species using the TaqMan real-time assay is shown in Table 3.23. All 25 isolates from species of C, D and F were correctly identified, including those epidemic serotypes Ad8, Ad19, Ad20 and Ad37 (species D). Those of species C identified were Ad1, Ad2, Ad5, and Ad6. Overall, adenovirus serotypes of 32_clinical isolates were identified correctly. In addition, two of six Ad7 belonging to species B were positive by the TaqMan assay. There were no false negative results observed with TaqMan system. When the PCR products of different Ad strains were analyzed by gel electrophoresis, they produced a specific band of the expected size without any primer dimers (data not shown).

Table 3.23

Comparison of TaqMan PCR assays with representative serotypes from different species of previously typed isolates

Species	Serotype	Hexon region TaqMan	Fiber Gene TaqMan
A	31	-	-
B	3	-	-
	7(6)[a]	-(4/6)[b]	- (6/6)
	11	-	-
	34	-	-
C	1(2)	+(2)	- (2/2)
	2(4)	+(4)	- (4/4)
	5	+	-
	6	+	-
D	8(3)	+(3)	- (3/3)
	19(3)	+(3)	- (3/3)
	20	+	-
	23	+	-
	37(2)	+ (2)	- (2/2)
E	4(2)	- (2)	-
F	40(7)	+(7)	+ (7)
	41(7)	+(7)	+ (7)

[a] Number in parenthesis shows number of samples tested.
[b] Number in parenthesis shows number of samples positive or negative.

Evaluation of the hexon real-time PCR primers in a standard PCR

The hexon primers for adenoviruses were evaluated in the standard PCR assay as described in Methods and Materials to establish sensitivity of the method. For this evaluation dilutions of Ad2 or Ad41 were added to aliquots of mock source water concentrates prepared by the same method as the water samples collected at the utilities. DNA was extracted from 100 µl samples with the developed GuSCN lysis buffer and Qiagen mini columns. The sensitivity of the assay for detection of Ad2 and Ad41 in water concentrates is shown in Figure 3.30.

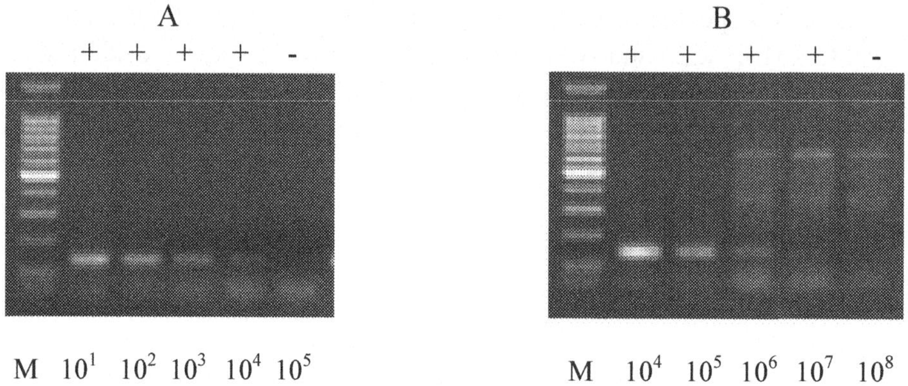

Figure 3.30. Sensitivity of primers for detection of Ad2 and Ad41 in lake water concentrates. Indicated dilutions of stock virus, Ad41 (A) or Ad2 (B) were spiked into aliquots of water concentrate prior to DNA extraction and tested in PCR.

Ad41 was detectable at dilutions yielding sensitivity of about 0.2 IU (a dilution of 10^{6}) and Ad2 at a sensitivity of 2 IU per assay (assay of 10^7 dilution). The sensitivity for absolute number of particles or genome copies may not be that dissimilar, because the particle to infectivity ratio of Ad41 from cell culture is higher than Ad2 (Mautner 1999). Basing the sensitivity on infectious units therefore yields a greater difference between the two types of Ad than might be observed if genome copies were evaluated. The particle to infectivity ratio of Ad41 in some studies is as high as 1000:1, whereas for Ad2 the ratio is 10:1. Assuming similar particle to infectivity ratios of our stock virus preparations, the detection limit per PCR of water concentrate should be about 20 particles or genomic copies.

This trial of the hexon primers for detection of Ad2 and Ad41 in water sample concentrates was successful. The sensitivity obtained with these primers is similar to that obtained with the use of published hexon primers with a hemi-nested assay (as described in Table 3.13). Therefore these primers were chosen for the screening of all utility water samples for detection of Ad. The results of those analyses will be presented in the next section of this report.

A real-time quantitative PCR was developed for human adenoviruses in clinical and environmental samples. The accuracy and the reproducibility of the real-time Ad PCR were confirmed by coded panel evaluation of 44 typed clinical isolates from which DNA was extracted. All adenovirus serotypes that belong to the different species of C, D and F were identified correctly. Those belonging to species F were identified by the fiber gene TaqMan PCR. The important epidemic keratoconjunctivitis-causing serotypes of species D (Ad8, Ad19 and Ad37), all serotypes of species C (Ad1, Ad2, Ad5 and Ad6) and both serotypes of species F (Ad40 and Ad41) were detected. Identification of adenovirus of species F (Ad 40 and Ad41) specifically is important since these are more commonly shed fecally and potentially found more often in the environment (Chapron et al. 2000). Whether the developed real-time PCR assay can detect animal adenoviruses was not tested. However, a GenBank search for compatibility of primers and probes indicated that the probe is not compatible with the published sequences of animal adenoviruses. Further work is needed to evaluate this possibility. The real-time PCR for Ad developed in this study has the potential to be used for rapid detection of human adenoviruses in both clinical and environmental samples, including water samples. A rapid specific test is needed for Ad detection in water because these viruses are included in the drinking water Contaminant Candidate List (CCL), and because several studies have reported adenovirus occurrence in environmental samples from diverse geographic locations (Puig et al. 1994; Jiang et al. 2001; Cho et al. 2000; Pina et al. 1998; Lee and Kim 2002). In addition, adenoviruses have been shown to be resistant to drinking water treatment (Gerba et al. 2002).

In summary, the real-time (RT-)PCR assays developed for the four different viruses provide a method for rapid screening of water concentrates for presence of viral nucleic acid indicating that viral contamination has occurred. After screening, the nucleic acid positive samples could be assayed for infectivity by standard methods or newly developed methods that we have described. More importantly, the real-time assays can also be used in conjunction with several other methods for the detection of infectious virus. These methods are not presented in this report; however, we have recently done some preliminary work with some of these approaches. For example, the adenovirus real-time PCR could be used for detection of mRNA in infected cells with the appropriate use of the forward and reverse primers. This would provide a rapid a quantitative approach without performing a nested reaction. Another approach could be used for enteroviruses and HAV. After analysis of the inoculum by real-time for quantitation of the input viral nucleic acid, real time RT-PCR analysis of HAV and enterovirus infected cells could be used to establish that viral replication occurred with the increase in quantity of positive viral strands in the cells. We have begun to explore this method for some wild-type HAV strains with good results. Finally, a very rapid detection method could be established with the use of the receptor-mediated capture of potentially infectious virus followed by real-time RT-PCR instead of conventional RT-PCR. The sCAR capture method we have developed for capture of CVB, could be followed by detection with real-time RT-PCR assay for enteroviruses. This would provide a rapid method for the detection of infectious virus within one day.

EVALUATION OF WATER CONCENTRATES FOR PRESENCE OF VIRUSES

General approach

As a first step in screening samples for virus presence, aliquots of the sample concentrates were directly analyzed for the target viruses (enteroviruses, adenoviruses, and Noroviruses) by conventional (RT-)PCR methods. Any samples found positive in the screening assays were further analyzed for potentially infectious viruses using the newly developed methods of this study, or analyzed in the newly developed real-time assays. An overview of the preparation of source and finished drinking water samples from collection to nucleic acid extraction is shown in Figure 3.31 (A and B).

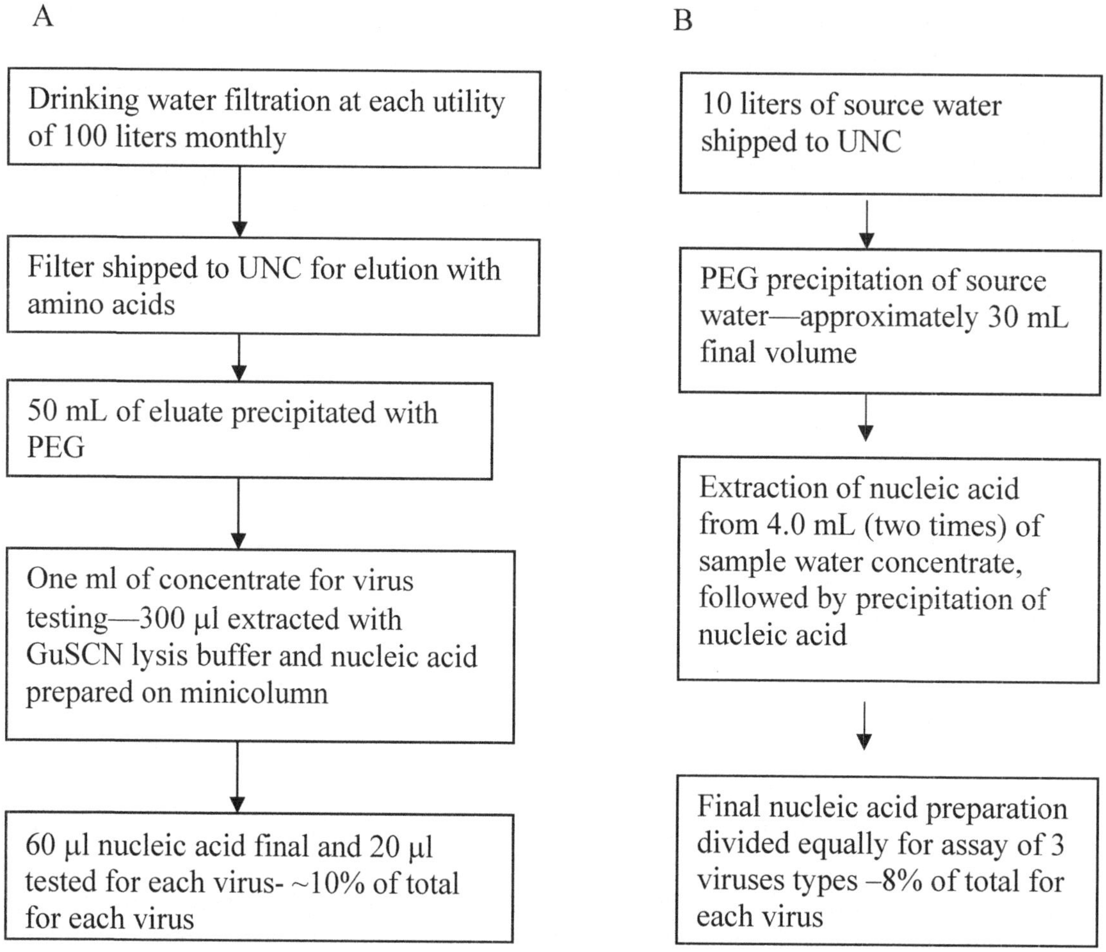

Figure 3.31. Collection, concentration, and analysis of drinking and source water samples—from utility to nucleic acid. A. Drinking water B. Source water (See Figure 2.1 for more details of preparation of nucleic acid from source water and Chapter 2 for preparation of nucleic acid from drinking water sample concentrates)

A total of eight mL of source water sample concentrate was prepared on two four-mL columns and divided for RNA precipitation for each of three viruses. Thus, about 8% of a sample was tested for each virus, and therefore, about 24% of each source water was tested for viruses in the screening assay by standard RT-PCR or PCR assays. Of the one-mL of drinking water sample concentrate, 300 µl was used to produce a nucleic acid sample of 50 µl, which was divided equally for testing of the viruses, resulting in about 10% of each drinking water sample tested for each virus. Remaining concentrates were retained for testing in appropriate newly developed assays if any samples were positive in the PCR and RT-PCR screening assays for virus presence, or for repeating the assay if needed.

Assay for Adenoviruses

All 104 drinking water concentrates and 104 source water concentrates were tested for Ad with the standard PCR method for detection of Ad DNA described in Chapter 2 using the hexon region primers evaluated in real-time RT-PCR. The hexon primers described for real-time PCR detect species C (Ad 2 and 5), D, and F (Ad40 and 41) were shown to be specific and sensitive for adenovirus detection by conventional PCR amplification. The sensitivity was obtained in the trials with water sample concentrates spiked with known amounts of Ad 2 and Ad 41 that yielded sensitivities of 2 IU and 0.2 IU, respectively. With this level of sensitivity, none of the 104 samples of source water concentrates yielded positives. Although some of samples had amplicons possibly the correct size on gel electrophoresis, sequence analysis was negative for Ad. This indicates that nonspecific DNA in water sample concentrates may contribute to non-specific amplicon production, and any amplicons must be verified by hybridization and or sequence analysis. All 104 drinking water concentrates also were negative for Adenovirus.

Assay for Noroviruses

All 104 samples of drinking water concentrate and all 104 samples of source water concentrates were tested for Noroviruses with previously described primers and RT-PCR assays, including a nested assay (Vinje´, Hamidjaja, Meschke, et al., 2001:Vinje et al. in preparation).

One source water sample was positive by RT-PCR. When it was sequenced and compared to the Norovirus nucleotide sequence database, it was identified as Hawaii (HV, G II.1). Nucleotide sequencing of the amplicons was done at the UNC Core Facility with Big Dye Terminator on a Perkin Elmer 3100 automated sequencer. This sample was not tested in the real-time assay since the developed assay is for GI isolates and not for GII.

All 104 drinking water samples were assayed by RT-PCR and were negative.

Assay for enteroviruses

All 104 samples concentrates of source water were tested for presence of enterovirus genomes with the standard RT-PCR assay and the panenterovirus 5'NTR primers described in Methods and Materials. No source water samples tested positive for enteroviruses by this RT-PCR assay. The sample size tested represented about 8% of the total sample and sensitivity of this assay is 0.1 PFU CVB3 per reaction. Therefore the concentration of viruses detectable by these primers (the lower detection limit) should be ~2 PFU per liter.

On initial first round screening of 104 drinking water sample concentrates, five samples produced amplicons of the correct size visualized on agarose gels stained with ethidium bromide. Two of these samples were also positive in a real-time TaqMan RT-PCR that included the specificity of a hybridization probe. These two samples from the standard RT-PCR assay were also sequenced. This was done on an amplicon obtained with a hemi-nested PCR of the first round RT-PCR products using the reverse of the panenterovirus primer pair and the forward primer used in the real-time RT-PCR for enteroviruses (see Table 2.2). Sequencing was performed on a 3100 Perkin Elmer automated sequencer with Big Dye terminator reactions. Sequences were subjected to BLAST analysis and were compared to GenBank sequences. Two different isolates were identified as Coxsackie A20. Distances of the neighbor-joining method (Fig. 3.32.) were calculated by the Kimura 2p method, available in MEGA, version 2.1.

Although each sample is identified as Coxsackie A20, they differ from each other by one nucleotide, and both are very different from the CVB3 laboratory strain of virus. When tested on the RF assay for infectious virus they were negative. Since theses viruses were obtained from a drinking water source this is not unexpected. Standard disinfection treatment at the water utility at the time would likely have inactivated the virus infectivity although possibly leaving the virus RNA detectable by RT-PCR

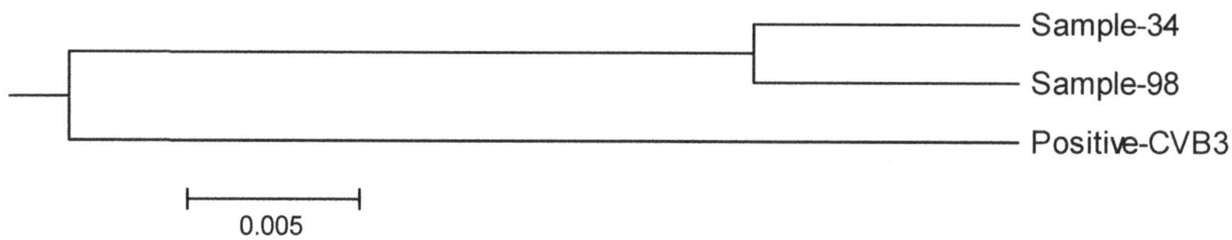

Figure 3.32. Phylogenetic tree constructed based on analysis of a 67-bp region of the 5'NTR of two isolates obtained in this study. CVB3 (Nancy), the laboratory control was also sequenced as a positive control for comparison.

With these findings for three virus groups, adenoviruses, Noroviruses, and enteroviruses, it was decided not to examine all 208 samples for HAV. This is because HAV is far less likely to be found than the above 3 virus types, based on its relative prevalence in and excretion by the population. Furthermore, a low level of HAV from a water sample by direct RT-PCR would not be positive in the assay for infectivity, given the inefficiency of HAV infectivity in cell cultures. We demonstrated that a higher titer was required at least for one HAV strain in order to replicate and produce RF indicative of infectious virus.

Summary

In summary, three virus positive samples were found when standard molecular methods were used to screen the archived drinking water and source water samples for the presence of viruses (infectious or non-infectious). One Norovirus and two enteroviruses were confirmed by sequence analysis. Unfortunately too few viruses were found to adequately evaluate the newly developed molecular methods for infectious virus detection. However, evaluation of the newly developed methods for detection of infectious viruses in water samples concentrates indicates that the methods will be useful in detection of infectious viruses. The standard (RT-)PCR detection methods used for screening all water samples were evaluated for detection sensitivity of inoculated virus in concentrates of sample source water including river, creek and pond. Similar end-point detection was demonstrated when RNA was extracted from water sample concentrates or from a laboratory buffer. This data indicates that if viruses were present at very low levels such as ~2 infectious units /liter in the case of enteroviruses, they should have been detected even in the presence of inhibitors from the water concentrates.

The developed RNA extraction method facilitated examination of a larger portion of the final concentrate. For examination of water samples by molecular methods, the ability to examine a larger portion of the sample is very important. This is because the small sample volumes typically extracted for RNA provide inadequate sensitivity to detect low levels of viruses in water. The improved RNA extraction method for larger sample volumes improves detection sensitivity 4-fold. This method should be further evaluated by applying it to water samples concentrated by other methods than those applied in this study.

Confirmation of virus positive environmental samples detected by standard molecular methods is cumbersome but necessary. This is because false positive results may be obtained when only gel visualization of PCR products is the criterion of positivity because the nucleic acids of a variety of microorganisms present in water concentrates can be inadvertently amplified. A standard approach to confirmation is specific probe hybridization of the nucleic acid products from PCR. The real-time RT-PCR developed in this project was very useful for confirmation of enterovirus positives in water sample concentrates. The one tube approach that includes hybridization provides a high level of sensitivity and is much more rapid than standard methods of confirming PCR products by hybridization or sequencing.

CHAPTER 4
SUMMARY, CONCLUSIONS AND, RECOMMENDATIONS

SUMMARY AND CONCLUSIONS

In this report we describe several new methods for the detection of important pathogenic viruses potentially transmitted by water. The increasing need for the reuse of water and the inadvertent reuse of natural sources of water impacted by continued and growing point and non-point human waste discharges, make detection of viruses in source and finished water supplies an important need. Viruses are more resistant to conventional water treatment methods than bacteria, although bacteria are used as the indicators organisms of water quality. Molecular methods for the detection of the nucleic acids of viruses, especially PCR and RT-PCR, have gained prominence in the past decade, and have been used successfully by research laboratories to detect viruses in environmental samples. This technology continues to evolve rapidly, with advances in rapid and automated systems and commercial reagents becoming available. While nucleic acid detection methods will have a significant role in the future of water safety, one aspect of detection that is not addressed currently with these methods is the detection of non-infectious viruses.

We have developed several molecular methods that will detect potentially infectious virus much more rapidly than the current the standard methods. Furthermore, we have developed a real-time quantitative (RT-)PCR for HAV in addition to all the viruses on the CCL published by EPA. This method is far more automated than standard PCR and provides in 60-90 minutes the specificity of a standard PCR with confirmatory hybridization, which otherwise takes days to complete. The real-time quantitative method also can be applied to the detection of infectious viruses in inoculated cultures.

One approach developed in this study for the detection of infectious viruses is the detection of unique nucleic acids in inoculated cells that are specific to the replication of either positive-strand RNA viruses, a category including many viruses contaminating water, or DNA viruses, such as Adenoviruses. Upon infection of host cells, positive strand RNA viruses produce a negative strand that is paired with the positive strand in a double-stranded RNA (RF) in infected cells. Detection of RF in cells inoculated with a model enterovirus CVB3 was possible within 24 hours of inoculation of a cell culture flask with as few as 4 infectious units. Cells inoculated with CVB3 inactivated with free chlorine contained no RF RNA. With this RF assay, replication of wild-type enteroviruses isolated from sewage, CVB4 and Echo1, were also detectable in 24-36 hr after infection of cell culture flasks with 4 infectious units. The specific RF RNA remains detectable by the assay as long as the cell culture is intact and even after lysis of cells occurs due to full CPE. Replication of wild-type HAV was also detected with the method for one isolate of virus in one cell line. HAV replication is very slow and weeks to months can be required to obtain replication of a wild-type strain. The RF method demonstrated replication within two weeks of culture by one strain of virus with known infectivity for primates with an inoculum of between 10^2 and 10^5 primate infectious units. Continued passage of the virus in culture confirmed RF presence.

In this study methods were developed to detect Adenovirus mRNA by RT-PCR in material from inoculated cell cultures. Because only infectious viruses can produce mRNA, this assay is specific for infectious viruses. The method was effective, sensitive and rapid for detecting low levels of infectious Ad in water sample concentrates. The mRNA RT-PCR method clearly distinguished infectious adenoviruses from non-infectious adenoviruses that were inactivated by free chlorine and UV radiation disinfection. The assay sensitivity was as low as 5 IU of Adenovirus 41 in 3 days of culture, and as few as 2 IU Ad2 and 10 IU Ad 41 in water concentrates inoculated onto cells for 7 days. The developed mRNA RT-PCR method is particularly applicable to detection of infectious Adenoviruses 40 and 41, because these viruses do not produce clear CPE in cell cultures but they do produce readily detectable mRNA in infected cells.

Coxsackie B viruses, laboratory and field-isolated, can be captured on magnetic beads coated with a soluble cellular receptor. This reaction is dependent upon the presence of a reactive intact virus coat (capsid). Production of the soluble coxsackie-adenovirus receptor (sCAR) was established with yields sufficient from one culture flask to produce hundreds of test reactions. After capture of the virus on magnetic beads, the RNA is released chemically and detected specifically and semi-quantitatively by RT-PCR amplification. The method is specific for infectious viruses as demonstrated by the lack of detection of chlorine-disinfected viruses. As few as 1-10 infectious units of virus can be detected in the assay of up to a 1 mL sample. Wild-type CVB isolates inoculated into river and lake water sample concentrates could be captured on the magnetic beads and detected by RT-PCR at levels as low as 1 cell culture infectious unit.

Real-time quantitative RT-PCR or PCR was developed and evaluated to detect most of the human pathogenic viruses that may contaminate water including: enteroviruses, HAV, Noroviruses and Adenoviruses. More than one type of virus can be detected in a single assay by a multiplex approach. A method developed for the simultaneous detection of HAV and enteroviruses has a sensitivity of 8 plaque forming units in 4.0 mL of source water concentrate (river or lake). This method was used to validate two water concentrate samples that were positive for enteroviruses with a standard RT-PCR assay. The real-time assays for HAV and for CVB model enterovirus yielded sensitivity as low as 0.8 PFU or approximately 50-100 virus particles. Primers and probe for real-time detection of Genogroup I Noroviruses were developed and demonstrated to detect all strains of the Genogroup in clinical specimens. An adenovirus real-time PCR was developed that can detect all Ad species that may contaminate water, including Ad 2, 5, 40 and 41, with a sensitivity of 0.01 infectious units of Ad2 and 10 IU of Ad41.An assay for specific detection of Ad 40 and 41 also was developed. The real-time PCR or RT-PCR assay can be obtained in 60-90 minutes after nucleic acid extraction providing a rapid, sensitive and specific method for detection of multiple viruses in water.

An improved RNA extraction method was developed which permits testing of a larger volume of water sample concentrates for viruses. It is a simple column method that uses mostly commercially available kits and related materials, does not require the use of organic solvents and can be done with only a low speed centrifugation. Ten liters of source water was concentrated to 30 mL and RNA was extracted from four mL of this sample, in contrast to the usual 100 μl sample used for RNA extraction on standard commercially available mini columns. A buffer developed in this project was used in conjunction with a commercially available column to accomplish the increased recovery of RNA. The midi column commercial kit contains buffer designed to extract from only one mL of sample, therefore, the improved method provided a four-fold increase sample volume to process for viral RNA recovery.

RECOMMENDATIONS FOR FUTURE RESEARCH

The work of this report demonstrates that molecular methods can be developed to selectively detect only infectious viruses from different types of water concentrates and that rapid real-time quantitative molecular methods are applicable to detection of waterborne viruses. Each method has been validated for specific waterborne virus groups and for the detection of enteric viruses in water sample concentrates. Further work is needed to expand these methods to other viruses or to combine the developed methods in ways that provide more rapid detection of only infectious viruses. Some of the necessary areas of focus should include:

1. Evaluation of different cell lines than the BGM for culture of viruses from field samples. The RF detection method we describe for positive-strand RNA viruses is very specific and more rapid than appearance of viral CPE, however the limiting factor is the use of appropriate cell lines and appropriate virus-cell adsorption conditions. A study focused on evaluation of enteroviruses concentrated from a sewage source assumed to contain a wide variety of viruses, such a large US city, should be performed as a way to further validate the methods for field viruses. Sample concentrates should be evaluated for virus production on a variety of cell lines already known to be good candidates for culture of enteric viruses. We demonstrated that only one cell line of four was susceptible for a single stool isolate of HAV. Further work is needed to evaluate cell lines and conditions for the detection of infectious HAV. Use of the RF detection method or the quantitative RT-PCR would make this process more efficient and accurate. Because HAV remains detectable by RT-PCR for long periods of time in environmental sources, evaluation of environmental isolates for infectivity is most important.
2. Demonstration of the success of the sCAR receptor for capture of CVB opens up the potential for evaluation of capture of other viruses with the appropriate cellular receptor. With the use of a variety of solid phase capture systems coated with multiple cellular receptors for a variety of viruses, a format that can capture several viruses simultaneously could be developed. A soluble receptor for HAV has been identified and demonstrated to neutralize virus, and the cellular receptor(s) for Noroviruses are being identified. The information about theses virus receptors provides the information to establish a capture system for infectious HAV and Noroviruses.
3. While the real-time RT-PCR and PCR as developed does not distinguish infectious or non-infectious viruses, it can be applied to specifically capturing infectious virus using the developed methods of immuno- or receptor capture on magnetic beads. The real-time assay is a one-tube rapid assay with the specificity of hybridization included when the TaqMan assay is used. Future research is recommended to couple the real-time quantitative (RT-)PCR detection methods with the capture methods and also to detect replication of viruses in cell culture. The quantitative aspect of the real-time assay provides a simple, reliable method to confirm an increase in viral genomes over that found in water sample concentrates, and it can be integrated into cell-culture RT-PCR and PCR methods.
4. Additional research is needed to evaluate the methods for recovery and detection of infectious viruses by applying them to viruses that have been inactivated by agents other than free chlorine or low pressure UV radiation. Viruses inactivated by medium pressure UV disinfection, chloramines, chlorine dioxide, ozone and the proteolytic enzymes

associated with microbes in natural waters and biological treatment processes (both water and wastewater) also need to be evaluated for their detectability by the developed methods.

5. Although the molecular techniques we have developed in this report are not immediately available to most analytical laboratories of water utilities, they can be applied in the future. Techniques such as real-time quantitative (RT-)PCR are currently becoming routine in clinical laboratories. The hardware, reagents, and trained staff are becoming more readily available. New kits are becoming commercially available for performance of the new molecular methods in a straightforward, simple manner. After standardization of the methods by validation in inter-laboratory trials, it is possible for the methods to be transferred to water utility laboratories. Utilities serving large metropolitan areas have or are building Level 2 biosafety laboratories, which could also be used for the techniques that involve inoculation of cell cultures. As these facilities become available, the molecular methods we have described for detection of infectious viruses in inoculated cell cultures could be performed. Methods involving use of immunocapture or cell receptor capture can be performed easily if standardized reagents are available. The capture of infectious virus followed by rapid real-time molecular detection would provide a powerful tool for rapid, quantitative detection infectious viruses in water samples.

REFERENCES

All the Virology on the Web. http://www.virology.net/ D. Sander, 1995-2004.

Alvarez, M.E. and R.T. O'Brien. 1982. Effects of chlorine concentration on the structure of poliovirus. *Appl Environ Microbiol* 43:237-239.

Anderson, D.A., B.C.Ross, and S.A.Locarnini. 1988. Restricted replication of hepatitis A virus in cell culture: encapsidation of viral RNA depletes the pool of RNA available for replication. *J Virol* 62:4201-4206.

Andreoletti, L., D.Hober, P.Becquart, S.Belaich, M.C.Copin, V.Lambert, and P.Wattre. 1997. Experimental CVB3-induced chronic myocarditis in two murine strains: evidence of interrelationships between virus replication and myocardial damage in persistent cardiac infection. *J Med Virol* 52:206-214.

Battigelli, D.A., D.Lobe, M.D.Sobsey. 1993. Inactivation of Hepatitis A Virus and other enteric viruses in water by ultraviolet Light. *Water Science and Technol* 27(3-4):339-342.

Barton, D.J. and J.B.Flanegan. 1997. Synchronous replication of poliovirus RNA: initiation of negative-strand RNA synthesis requires the guanidine-inhibited activity of protein 2C. *J Virol* 71:8482-8489.

Bergelson, J.M., J.A.Cunningham, G.Droguett, E.A.Kurt-Jones, A.Krithivas, J.S.Hong, M.S. Horwitz, R.L.Crowell and R.W. Finberg. 1997. Isolation of a common receptor for Coxsackie B viruses and Adenoviruses 2 and 5. *Science* 275:1320-1323.

Boom, R., C.J.Sol, M.M.Salimans, C.L.Jansen, P.M.Wertheim-van Dillen, and J.van der Noordaa. 1990. Rapid and simple method for purification of nucleic acids. *J Clin Microbiol* 28:495-503.

Brandt, C.D., H. W. Kim, W. J. Rodriguez, J.O. Arrobio, B.C. Jeffries, E.P. Stallings, C. Lewis, A. J. Miles, M. K. Gardner, and R. H. Parrott. 1985. Adenoviruses and pediatric gastroenteritis. Journal of Infectious Diseases. 151(3):437-43.

Brown, M., H.L. Wilson-Friesen, and F. Doane. 1992. A block in release of progeny virus and a high particle-to-infectious unit ratio contribute to poor growth of enteric adenovirus types 40 and 41 in cell culture. *Journal of Virology* 66(5):3198-205.

Castignolles, N., F.Petit, I.Mendel, L.Simon, L.Cattolico, and C.Buffet-Janvresse. 1998. Detection of adenovirus in the waters of the Seine River estuary by nested-PCR. *Mol Cell Probes* 12:175-180.

Chapron, C.D., N.A.Ballester, J.H.Fontaine, C.N.Frades, and A.B.Margolin. 2000. Detection of astroviruses, enteroviruses, and adenovirus types 40 and 41 in surface waters collected and evaluated by the information collection rule and an integrated cell culture-nested PCR procedure. *Appl Environ Microbiol* 66:2520-2525.

Chaves, R.L., J.Graff, A.Normann, and B.Flehmig. 1994. Specific detection of minus strand hepatitis A virus RNA by Tail-PCR following reverse transcription. *Nucleic Acids Res* 22:1919-1920.

Cho, H.B., S.H.Lee, J.C.Cho, and S.J.Kim. 2000. Detection of adenoviruses and enteroviruses in tap water and river water by reverse transcription multiplex PCR. *Can J Microbiol* 46:417-424.

Chung, H., L.A. Jaykus, and M..D. Sobsey. 1996. Detection of human enteric viruses in oysters by in vivo and in vitro amplification of nucleic acids. *Appl Environ Microbiol* 61(10):3772-8.

Cohen, JI, J.R Ticehurst, R.H.Purcell, A.Buckler-White, B.M.Baroudy. 1987. Complete nucleotide sequence of wild-type hepatitis A virus: comparison with different strains of hepatitis A virus and other Picornaviruses. *J Virol* 61:50-9.

Costa-Mattioli, M., S.Monpoeho, E.Nicand, M.H.Aleman, S.Billaudel, and V.Ferre. 2002. Quantification and duration of viraemia during hepatitis A infection as determined by real-time RT-PCR. *J Viral Hepat* 9:101-106.

Craggs, J.K., J.K.Ball, B.J.Thomson, W.L.Irving, and A.M.Grabowska. 2001. Development of a strand-specific RT-PCR based assay to detect the replicative form of hepatitis C virus RNA. *J Virol Methods* 94:111-120.

Cromeans, T., H.A.Fields, and M.D.Sobsey. 1989. Replication kinetics and cytopathic effect of hepatitis A virus. *J Gen Virol* 70 (8):2051-2062.

Cromeans, T.L., O.V.Nainan, and H.S.Margolis. 1997. Detection of hepatitis A virus RNA in oyster meat. *Appl Environ Microbiol* 63:2460-2463.

Cromeans, T.L.; Sobsey, M.D.; Fields, H.A. 1987. Development of a plaque assay for a cytopathic rapidly replicating isolate of hepatitis A virus. J. Med. Virol. 22:45-56.

Cromeans, T.L., M.O.Favorov, O.V.Nainan, and H.S. Margolis. Hepatitis A and E viruses. In *Foodbourne Disease Handbook;* Hui, Y.H., S.A.Sattar, K.D.Murrell, W.Nip and P.S.Stanfield. Eds.; 2nd ed; Marcel Dekker, Inc. New York, NY, 2001, pp 23-76.

Cunningham, L., N.E.Bowles, R.J.Lane, V.Dubowitz, and L.C.Archard. 1990. Persistence of enteroviral RNA in chronic fatigue syndrome is associated with the abnormal production of equal amounts of positive and negative strands of enteroviral RNA. *J Gen Virol* 71 (Pt 6):1399-1402.

De Jong, J. C., A. G. Wermenbol, M. W. Verweij-Uijterwaal, K. W. Slaterus, P. Wertheim-Van Dillen, G. J. Van Doornum, S. H. Khoo, and J. C. Hierholzer. 1999. Adenoviruses from human immunodeficiency virus-infected individuals, including two strains that represent new candidate serotypes Ad50 and Ad51 of species B1 and D, respectively. *Journal of Clin Microbiol.* 37(12)**:**3940-5.

De Serres, G., T.L.Cromeans, B.Levesque, N.Brassard, C.Barthe, M.Dionne, H.Prud'homme, D.Paradis, C.N.Shapiro, O.V.Nainan, and H.S.Margolis. 1999. Molecular confirmation of hepatitis A virus from well water: epidemiology and public health implications. *J Infect Dis* 179:37-43.

Divizia, M., V. Ruscio, A.M. Degener, and A. Pana. 1998. Hepatitis A virus detection in wastewater by PCR and hybridization. *Microbiol Rev* 21:161-167.

Feigelstock, D., P. Thompson, P. Mattoo, and G. G. Kaplan. 1998. Polymorphisms of the hepatitis A virus cellular receptor 1 in African Green monkey kidney cells results in antigenic variants that do not react with protective monoclonal antibody 190/4. *J. Virol.* 72:6218-6222.

Gantzer, C., A. Maul, J.M.Audic, and L.Schwartzbrod. 1998. Detection of infectious enteroviruses, enterovirus genomes, somatic coliphages, and *Bacteroides fragilis* phages in treated wastewater. *Appl Environ Microbiol* 64:4307-4312.

Gerba, C.P., D.M.Gramos and N. Nwachuku. 2002. Comparative inactivation of Enteroviruses and Adenovirus 2 by UV light. *Appl Environ Microbiol* 68(10) 5167-5169.

Graham, F.L., J.Smiley, W.C.Russell, and R.Nairn. 1977. Characteristics of a human cell line transformed by DNA from human adenovirus type 5. *J Gen Virol* 36:59-74.

Green, D.H. and G.D.Lewis. 1995. Enzymatic amplification of enteric viruses from wastewaters. *Water Sci. Tech.* 31: 329-336.

Harakeh, M. and M.Butler. 1984. Inactivation of Human Rotavirus, SA11 and Other Enteric viruses in effluent by disinfectants. *J Hygiene* 84:247-55.
Hollinger, F.B. and S.V. Emerson. 2001. Hepatitis A Viruses, In: *Fields Virology*, 3rd Ed. Vol 1, D.M. Knipe and P.M. Howley, Eds. Lippincot, Williams, and Wilkins, Philadelphia, PA.
Ijzerman, M..M., D.R.Dahling, and G.S.Fout. 1997. A method to remove environmental inhibitors prior to the detection of waterborne enteric viruses by reverse transcription-polymerase chain reaction. *J Virol Methods* 63:145-153.
Jiang, S., R.Noble, and W.Chu. 2001. Human adenoviruses and coliphages in urban runoff-impacted coastal waters of Southern California. *Appl Environ Microbiol* 67:179-184.
Johnson, K.L. and P. Sarnow. Viral RNA synthesis. In *Human Enterovirus Infections;* Rotbart, H.A.; Ed. American Society for Microbiology, Washington D.C., 1995, pp 95-112.
Jothikumar, N., R.Paulmurugan, P.Padmanabhan, R.B.Sundari, S.Kamatchiammal, and K.S.Rao. 2000. Duplex RT-PCR for simultaneous detection of hepatitis A and hepatitis E virus isolated from drinking water samples. *J Environ Mon.* 2:587-590.
Khanna, B., J.E. Spelbring, B.L. Innis, and B.H. Robertson. 1992. Characterization of a genetic variant of human hepatitis A virus. *J Med Virol* 36(2):118-24
Krawczynski, N. R. Quadri, L. Rubbia-Brandt, M. Mondelli, J. Zarski, A. Hadengue. 1999. Detection of genomic and minus-strand of hepatitis C virus RNA in the liver of chronic hepatitis C patients by strand-specific semi-quantitative reverse-transcriptase polymerase chain reaction. *Hepatology* 29(2): 536-542.
Le Guyader, F., M.K.Estes, M.E.Hardy, F.H.Neill, J.Green, D.W.Brown, and R.L.Atmar. 1996. Evaluation of a degenerate primer for the PCR detection of human caliciviruses. *Arch Virol* 141:2225-2235.
Lanford, R.E., D.Chavez, F.V.Chisari, and C.Sureau. 1995. Lack of detection of negative-strand hepatitis C virus RNA in peripheral blood mononuclear cells and other extrahepatic tissues by the highly strand-specific rTth reverse transcriptase PCR. *J Virol* 69:8079-8083.
Laskus, T., M.Radkowski, L.F.Wang, H.Vargas, and J.Rakela. 1998. Detection of hepatitis G virus replication sites by using highly strand-specific Tth-based reverse transcriptase PCR. *J Virol* 72:3072-3075.
Lee, S.H. and S.J.Kim. 2002. Detection of infectious enteroviruses and adenoviruses in tap water in urban areas in Korea. *Water Res* 36:248-256.
Ma, J.-F., T.M. Straub, I.L.Pepper, and C.P. Gerba. 1994. Cell culture and PCR Determination of Poliovirus inactivation by disinfectants. *Appl Environ Microbiol* 60:4203-4206.
Mackay, I.M., K.E.Arden , A. Nitsche. 2002. Real-time PCR in virology. *Nucleic Acids Research* 30(6):1292-1305.
Maier, A., D. Tougianidou, A. Wiedenmann, K. Botzenhart. 1995. Detection of Poliovirus by cell culture and by PCR after UV Disinfection. *Water Science Tech* 31:141-145.
Margolis, H.S. and O.V.Nainan. 1990. Identification of virus components in circulating immune complexes isolated during hepatitis A virus infection. *Hepatology* 11:31-37.
Mautner, V. 1999. Methods for Growth and Purification of Enteric Adenovirus Type 40. In *Adenovirus Methods and Protocols*. Ed. By William S.M. Wold. Humana Press, Totowa, NJ. p 283.
McCaustland, K.A.; Bond, W.W.; Bradley, D.W.; Ebert, J.W.; Maynard, J.E. 1982. Survival of hepatitis A virus in feces after drying and storage for 1 month. *J. Clin. Microbiol.* 16:957-958.

Molecular Probes, Inc., Copyright 1996-2004, Invitrogen Life Technologies http://www.probes.com/handbook/figures/0710.html

Monpoeho, S., A.Dehee, B.Mignotte, L.Schwartzbrod, V..Marechal, J.C. Nicolas, JC *et al.* .2000. Quantification of enterovirus RNA in sludge samples using single tube real-time RT-PCR. *Biotechniques* 29(1):88-93.

Monpoeho, S., A.Maul, B.Mignotte-Cadiergues, L.Schwartzbrod, S.Billaudel, and V.Ferre. 2001. Best viral elution method available for quantification of enteroviruses in sludge by both cell culture and reverse transcription-PCR. *Appl Environ Microbiol* 67:2484-2488.

Nemerow, G.R. 2000. Cell receptors involved in adenovirus entry. *Virology* 274:1-4.

Novak, J.E. and K.Kirkegaard. 1991. Improved method for detecting poliovirus negative strands used to demonstrate specificity of positive-strand encapsidation and the ratio of positive to negative strands in infected cells. *J Virol* 65:3384-3387.

Oberste, M.S., K.Maher, D.R.Kilpatrick, M.R.Flemister, B.A.Brown, and M.A.Pallansch. 1999A. Typing of human enteroviruses by partial sequencing of VP1. *J Clin Microbiol* 37:1288-1293.

Oberste, M.S., K.Maher, D.R.Kilpatrick, and M.A.Pallansch. 1999B. Molecular evolution of the human enteroviruses: correlation of serotype with VP1 sequence and application to picornavirus classification. *J Virol* 73:1941-1948.

Oberste, M.S., K.Maher, M.R.Flemister, G.Marchetti, D.R.Kilpatrick, and M.A.Pallansch. 2000. Comparison of classic and molecular approaches for the identification of untypeable enteroviruses. *J Clin Microbiol* 38:1170-1174.

Papapetropoulou, M. and A.C.Vantarakis. 1998. Detection of adenovirus outbreak at a municipal swimming pool by nested PCR amplification. *J Infect* 36:101-103.

Pina, S., M.Puig, F.Lucena, J.Jofre, and R.Girones. 1998. Viral pollution in the environment and in shellfish: human adenovirus detection by PCR as an index of human viruses. *Appl Environ Microbiol* 64:3376-3382.

Puig, M., J.Jofre, F.Lucena, A.Allard, G.Wadell, and R.Girones. 1994. Detection of adenoviruses and enteroviruses in polluted waters by nested PCR amplification. *Appl Environ Microbiol* 60:2963-2970.

Racaniello,V.R. 1996. Early events in poliovirus infection: virus-receptor interactions. *Proc Natl Acad Sci U S A* 93(21):11378-11381.

Read, S.J., J.L.Mitchell, and C.G.Fink. 2001. LightCycler multiplex PCR for the laboratory diagnosis of common viral infections of the central nervous system. *J Clin Microbiol* 39:3056-3059.

Reetoo, K.N., S.A.Osman, S.J.Illavia, C.L.Cameron-Wilson, J.E.Banatvala, and P.Muir. 2000. Quantitative analysis of viral RNA kinetics in coxsackievirus B3-induced murine myocarditis: biphasic pattern of clearance following acute infection, with persistence of residual viral RNA throughout and beyond the inflammatory phase of disease. *J Gen Virol* 81:2755-2762.

Reynolds, K.A., C.P.Gerba, and I.L.Pepper. 1996. Detection of infectious enteroviruses by an integrated cell culture-PCR procedure. *Appl Environ Microbiol* 62:1424-1427.

Roelvink, P.W., A.Lizonova, J.G.Lee, Y.Li, J.M.Bergelson, R.W.Finberg, D.E.Brough, I.Kovesdi, and T.J.Wickham. 1998. The coxsackievirus-adenovirus receptor protein can function as a cellular attachment protein for adenovirus serotypes from subgroups A, C, D, E, and F. *J Virol* 72:7909-7915.

Rosenfield, S.I and L.A. Jaykus. 1999. A multiplex reverse transcriptase polymerase chain reaction method for the detection of foodborne viruses. *J Food Prot* 62:1210-1214.

Rossmann, MG, He, Y, Kuhn RJ. 2002. Picornavirus-receptor interactions. *Trends in Microbiology* 10(7):324-331.

Schmitz, G. and A.Dotzauer. 1998. Proof of hepatitis A virus negative-sense RNA by RNA/DNA-hybrid detection: a method for specific detection of both viral negative- and positive-strand RNA species. *Nucleic Acids Res* 26:5230-5232.

Schwab, K.J., R.De Leon, and M.D.Sobsey. 1995. Concentration and purification of beef extract mock eluates from water samples for the detection of enteroviruses, hepatitis A virus, and Norwalk virus by reverse transcription-PCR. *Appl Environ Microbiol* 61:531-537.

Shin, G.-A. and M.D. Sobsey. 1998. Reduction of Norwalk virus, poliovirus 1 and coliphage MS2 by free chlorine, chlorine dioxide and ozone disinfection of water. Proc. 1998. Water Quality Technology Conference, Amer. Water Works Assoc. Denver, CO .

Sobsey, M.D., T.Fuji and R.M. Hall. 1991. Inactivation of cell-associated and dispersed hepatitis A virus in water. *J. American Water Works Assoc* 83: 64-67.

Sobsey, M.D., D. Battigelli, G. Shin, and S. Newland. 1998A. RT-PCR underestimation of reductions of poliovirus 1 and coliphage MS2 by free chlorine, chlorine dioxide, and UV disinfection of water. *Water Sciences and Technology* 38(12):91-94.

Sobsey, M.D., M.J. Casteel, H. Chung, G. Lovelace, O.D. Simmons, III, and J.S. Meschke. 1998B. Innovative Technologies for Wastewater Disinfection and Pathogen Detection. pp. 483-493. In *Disinfection 1998*. The Latest Trends in Wastewater Disinfection: Chlorination vs. UV Disinfection. Water Environment Federation, Alexandria, VA.

Staub, U., H.Polivka, and H.J.Gross. 1995. Two rapid microscale procedures for isolation of total RNA from leaves rich in polyphenols and polysaccharides: application for sensitive detection of grapevine viroids. *J Virol Methods* 52:209-218.

Thurman, R.B. and C.P. Gerba. 1988. Molecular mechanisms of viral inactivation by water disinfectants. In *Advances in Applied Microbiology*. Vol.33Academic Press. Pp. 75-105.

Tiemessen, C.T. and A.H. Kidd. 1996. Adenovirus type 40 and 41 replication: cell-related differences in viral gene transcription. *Mol cell Probes* 10(4):279-287.

Tiemessen, C.T., M.J.Nel, and A.H.Kidd. 1996. Adenovirus 41 replication: cell-related differences in viral gene transcription. *Mol Cell Probes* 10:279-287.

Verstrepen, W.A., S.Kuhn, M.M.Kockx, M.E.Van De Vyvere, and A.H.Mertens. 2001. Rapid detection of enterovirus RNA in cerebrospinal fluid specimens with a novel single-tube real-time reverse transcription-PCR assay. *J Clin Microbiol* 39:4093-4096.

Vinjé, J., R.A. Hamidjaja, J.S. Meschke, E-J. Collins-McNeal, C.L. Moe, and M.D. Sobsey. 2001. Evaluation of Different RT-PCR primer pairs for the detection of low numbers of Norwalk-like viruses as reference. 101st General Meeting of the American Society for Microbiology, May 20-24, Orlando, FL.

Vinjé, J., R. A. Hamidjaja , and M. D. Sobsey. (in-preparation) Development and application of a capsid VP1 (region D) based reverse transcription PCR assay for genotyping of genogroup I and II noroviruses. J Virol Methods. Volume 116, Issue 2, Pages 109-214 (15 March 2004)

U.S. EPA. 1998. "Federal Register notice about the Contaminant Candidate List" 63 FR No. 193 10273, March 2.

Wadell, G. 1984. Molecular epidemiology of human adenoviruses. *Curr Top Microbiol Immunol* 110:191-220.

Watkins-Riedel, T., M.Woegerbauer, D.Hollemann, and P.Hufnagl. 2002. Rapid diagnosis of enterovirus infections by real-time PCR on the LightCycler using the TaqMan format. *Diagn Microbiol Infect Dis* 42:99-105.

Weston, P.D. and S.Avrameas. 1971. Proteins coupled to polyacrylamide beads using glutaraldehyde. *Biochem Biophys Res Commun* 45:1574-1580.

Xu, W., M.C.McDonough, and D.D.Erdman. 2000. Species-specific identification of human Adenoviruses by a multiplex PCR assay. *J Clin Microbiol* 38:4114-4120.

Xu, W. and D.D.Erdman. 2001. Type-specific identification of human Adenovirus 3, 7, and 21 by a multiplex PCR assay. *J Med Virol* 64:537-542.

Yamamoto, K.R., B.M.Alberts, R.Benzinger, L.Lawhorne, and G.Treiber. 1970. Rapid bacteriophage sedimentation in the presence of polyethylene glycol and its application to large-scale virus purification. *Virology* 40:734-744.

Yuen, L.K., M.G.Catton, B.J.Cox, P.J.Wright, and J.A.Marshall. 2001. Hemi-nested multiplex reverse transcription-PCR for detection and differentiation of Norwalk-like virus genogroups 1 and 2 in fecal samples. *J Clin Microbiol* 39:2690-2694.

ABBREVIATIONS

A549	epithelial lung carcinoma cell line
Ab	antibody
Ad	Adenovirus(es)
Ad2	Adenovirus subtype 2
AMV RT	Reverse Transcriptase from avian myeloblastosis virus
ATCC	American Type Culture Collection
AwwaRF	Awwa Research Foundation
BE	beef extract
bp	base pair(s)
BGM	Buffalo Green Monkey cell line
BHQ	Black Hole Quenchers™
BSA	bovine serum albumin
Caco2	human colonic carcinoma cell line
CCL	candidate contaminant list
CDC	Centers for Disease Control
cDNA	complementary deoxyribonucleic acid
CVB	Coxsackievirus(s)
CVB3	coxsackievirus B3 strain
CVB4	coxsackievirus B4 strain
cm	centimeter
cm^2	centimeter squared
CPE	cytopathic (cytopathogenic) effects
Ct	contact time (mg min/L)
DPBS	Dulbecco's phosphate buffered saline
DNA	deoxyribonucleic acid
DNase	deoxyribonuclease
dNTP	deoxynucleotide triphosphate or deoxynucleotide
dsRNA	double-stranded ribonucleic acid
E1A	early antigen of Adenovirus
EDTA	ethylenediaminetetraacetic acid
EMEM	essential modified eagle's medium
EtOH	ethanol
EV	enterovirus
FAM	fluorescent dye
FBS	fetal bovine serum
FLAG	an 8-aminoacid hydrophilic peptide
FRhK	fetal rhesus kidney cell line
FRET	fluorescence resonance energy transfer

G293	Graham cells-human embryonic cell line
g	gravity
G	gram, glycine, genogroup, or guanine
GCG	Genetics Computer Group
GuSCN	guanidine thiocyanate
HAV	hepatitis A virus
HEPES	hydroxymethylpiperazine ethanesulfonic acid
Hex	hexon
HLD-2	hepatitis A virus strain, genotype 1A
HM-175	hepatitis A virus strain, genotype 1B
hrs	hours
HCl	hydrochloric acid
IC-RT-PCR	immunocapture reverse transcriptase polymerase chain reaction
IgG	immunoglobulin G
IU	infectious units
Inoc.	Inoculum
K	thousand
kG	kilograms
L	liter
M	molecular weight marker DNA or moles (molar)
MA-104	rhesus monkey kidney
MEM	minimum essential medium
mG	milligram
min	minutes
mL	milliliter
mM	millimole(s) or millimolar
MOI	multiplicity of infection
MPN	most probable number
μG	microgram
μL	microliter
μM	micromolar
N	number or negative control
NaCl	sodium chloride
NaOH	sodium hydroxide
NCR	non-coding region
ng	nanogram
nt	nucleotide(s)
NT	not tested
NTR	non-translated region
NV	Norovirus (formerly Norwalk-like virus or NLV)

PBS	phosphate buffered saline
PCR	polymerase chain reaction
PEG	polyethylene glycol
PFU	plaque-forming units
pH	log hydrogen ion concentration
PI	post infection
PV1	poliovirus type 1
RF	replicative form
RNA	ribonucleic acid
RNase	ribonuclease
ROX	fluorescent dye
RT	reverse transcription
RTase	reverse transcriptase
(RT-)PCR	reverse transcription-polymerase chain reaction or polymerase chain reaction
sCAR	soluble coxsackie-adenovirus receptor
ssDNA	single stranded deoxyribonucleic acid (ssDNA)
ssRNA	single-stranded ribonucleic acid (ssRNA)
Sup't	supernatant
TAE buffer	Tris-acetate-EDTA buffer
Taq	Thermus aquaticus
$TCID_{50}$	50% tissue culture infectious dose
T_m	melting temperature
TN5-B14	insect cell line
TNT	Tris, NaCL, and Tween
Tris	tris (hydroxymethyl) aminomethane
USEPA	United States Environmental Protection Agency
UV	ultraviolet
V/cm	volts per centimeter
VP1	viral polypeptide 1
v/v	volume per volume
w/v	weight per volume
X	times

IWA Publishing's authorised EU representative for General Product Safety Regulations is Diane D'Arras, 15 rue Duret, 75116 Paris, France, e-mail: safety@iwap.co.uk.

Printed and bound by CPI Group (UK) Ltd, Croydon, CR0 4YY
17/09/2026
02218810-0014